T0234702

SpringerBriefs in Applied Sciences and Technology

SpringerBriefs present concise summaries of cutting-edge research and practical applications across a wide spectrum of fields. Featuring compact volumes of 50 to 125 pages, the series covers a range of content from professional to academic.

Typical publications can be:

- A timely report of state-of-the art methods
- An introduction to or a manual for the application of mathematical or computer techniques
- A bridge between new research results, as published in journal articles
- A snapshot of a hot or emerging topic
- An in-depth case study
- A presentation of core concepts that students must understand in order to make independent contributions

SpringerBriefs are characterized by fast, global electronic dissemination, standard publishing contracts, standardized manuscript preparation and formatting guidelines, and expedited production schedules.

On the one hand, **SpringerBriefs in Applied Sciences and Technology** are devoted to the publication of fundamentals and applications within the different classical engineering disciplines as well as in interdisciplinary fields that recently emerged between these areas. On the other hand, as the boundary separating fundamental research and applied technology is more and more dissolving, this series is particularly open to trans-disciplinary topics between fundamental science and engineering.

Indexed by EI-Compendex, SCOPUS and Springerlink.

More information about this series at http://www.springer.com/series/8884

Martin Thomas Horsch · Silvia Chiacchiera ·
Welchy Leite Cavalcanti · Björn Schembera

Data Technology in Materials Modelling

 Springer

Martin Thomas Horsch [ID]
High-Performance Computing
Center Stuttgart
Stuttgart, Germany

Welchy Leite Cavalcanti
Fraunhofer IFAM
Bremen, Germany

Silvia Chiacchiera [ID]
STFC Daresbury Laboratory
Daresbury, Cheshire, UK

Björn Schembera [ID]
High-Performance Computing
Center Stuttgart
Stuttgart, Germany

ISSN 2191-530X   ISSN 2191-5318   (electronic)
SpringerBriefs in Applied Sciences and Technology
ISBN 978-3-030-68596-6   ISBN 978-3-030-68597-3   (eBook)
https://doi.org/10.1007/978-3-030-68597-3

This Springer imprint is published by the registered company Springer Nature Switzerland AG
The registered company address is: Gewerbestrasse 11, 6330 Cham, Switzerland

Preface

With this Springer Brief volume, we aim at disseminating information on recent work, including work in progress, towards developing metadata standards and digital platforms in the field of materials modelling. This work targets applications in the engineering sciences as well as industrial research and development in chemical and mechanical process technology. Digitalization in these fields is significantly advanced by public funding, notably through EU's Horizon 2020 and (prospectively) Horizon Europe programmes, the DFG programme for National Research Data Infrastructures (NFDI) and BMBF programmes including Material Digital, among many others.

The viability and uptake of metadata standards and digital platforms depend strongly on an exchange of ideas between development efforts. In our field, community organizations such as the Research Data Alliance, the European Materials Modelling Council and CECAM (Centre Européen de Calcul Atomique et Moléculaire) support this exchange through a multitude of events, white papers, and coordinated activities and task groups. It is our hope that all these endeavours will collaborate as closely as possible and jointly be successful at making FAIR data management ubiquitous in materials modelling. This work is intended as an input to this collaboration that is necessary, in our view. It is therefore intentionally published at a stage long before these developments can be regarded as finished, and we would like to encourage its readers to take this as an opportunity to engage in criticism and formulate recommendations for future work.

Daresbury, UK   Martin Thomas Horsch
Daresbury, UK   Silvia Chiacchiera
Bremen, Germany   Welchy Leite Cavalcanti
Stuttgart, Germany   Björn Schembera
September 2020

Acknowledgements

The co-authors M. T. H., S. C. and W. L. C. acknowledge funding from EU's Horizon 2020 research and innovation programme under grant agreement no. 760907 (VIMMP). The co-author B. S. acknowledges funding from the German Federal Ministry of Education and Research (BMBF) under grant no. FDM-008 (Dipl-Ing). This work was facilitated by activities of Inprodat e.V., Kaiserslautern.

The authors would like to thank N. Adamovic, B. Andreon, P. Asinari, F. Bach, Y. Bami, E. Bayro Kaiser, M. F. Bernreuther, A. Bhave, G. Boccardo, T. Bönisch, H. Brüning, V. Bunakov, P. Carbone, V. Carrillo Beber, M. Chiricotto, J. Díaz Brañas, S. Druskat, F. Duchateau, A. Duff, A. M. Elena, J. D. Elliott, E. Fayolle, A. Fiseni, Y. Fournier, J. Francisco Morgado, J. Friis, E. Ghedini, G. Goldbeck, P. Gralka, J. Handsel, N. Hansen, A. Hashibon, C. Henry, M. Hoffmann, D. Iglezakis, E.-D. Janotka, G. Kanagalingam, P. Klein, R. Kodžius, N. A. Konchakova, R. Kunze, V. Kushnarenko, M. Lísal, A. Lister, V. Lobaskin, S. Metz, J.-P. Minier, G. Mogni, C. Niethammer, P. Neumann, P. Noyret, E. Ntainjua, I. Pagonabarraga Mora, S. Pařez, T. Petrenko, B. Planková, G. Reina, M. M. Resch, P. Schiffels, G. J. Schmitz, A. Scotto di Minico, M. A. Seaton, K. Sen, A. Simperler, K. Šindelka, A. Singhal, S. Stephan, V. Stobiac, G. Summer, I. T. Todorov, D. Toti, M. Verlage, J. Vrabec, and C. Yong for fruitful discussions at various stages of this work.

Contents

Chapter 1
Introduction

1.1 Digitalization and Data Management

Digitalization is one of the driving forces of technological and social progress today. In the engineering sciences, in combination with a great variety of quantitatively reliable modelling and simulation approaches, digitalization supports the development of what has become known as Industry 4.0 by contributing to virtual manufacturing through cyber-physical systems. To predict thermodynamic, mechanical and other physical properties of materials and processes, data-driven and physics-based models are combined [1], supported by massively parallel simulation methods that continue to become more scalable and performant [2]; model databases are developed [3, 4], and data with a heterogeneous provenance (i.e. origin), based on different methods and coming from different sources [5, 6], are integrated into shared data infrastructures [7]. A multitude of names have been proposed for the related lines of work in academic and industrial research and development, including Integrated Computational Materials Engineering (ICME), with a focus on solids [8, 9], Computational Molecular Engineering (CME), with a focus on fluids [4, 6] and process data technology or computer-aided process engineering (CAPE), with an orientation towards process technology and CAPE-OPEN-based simulation technology [10–12]. This book discusses data management in materials modelling, which is here understood to encompass all these fields.

Digitalization is achieved in two steps: First, data must be available in digital form. The process of making data available digitally is referred to as digitization; in the engineering sciences, with certain exceptions (e.g. data published in old volumes of journals that have not yet been digitized by scanning), this can usually be presupposed. However, the possible use of raw unannotated digital data, also known as dark data [13, 14], is very limited. Beyond digitization, a second step is therefore required for digitalization to ensure that the data are and remain findable, accessible, interoperable and reusable (FAIR): These are the FAIR principles of data management or data stewardship [15–17]. For some applications, such as mediation systems [18, 19] for Ontology-Based Data Access (OBDA) to distributed

© The Author(s) 2021
M. Horsch et al., *Data Technology in Materials Modelling*,
SpringerBriefs in Applied Sciences and Technology,
https://doi.org/10.1007/978-3-030-68597-3_1

1

heterogeneous data sources [20, 21], these four principles are jointly fundamental and cannot be separated from each other. In other typical cases, e.g. for complex simulation workflows, interoperability is the main concern [22, 23]; however, even in these cases, it is reasonable to follow good practices concerning all the aspects of FAIR data management. Findability and accessibility are supported by systems of persistent identifiers, with Digital Object Identifiers (DOIs) now covering almost all scientific publications, as well as platforms and legal solutions for open-access publishing.

The single aspect of greatest importance to the findability, interoperability and reusability of data is semantically characterized data annotation, i.e. the provision of metadata in a way that is widely agreed and understood on the basis of community-governed metadata standardization. This is the main topic of this book, where the focus will be on the interoperability aspects of FAIR data management and its practical realization by digital platforms and data infrastructures for materials modelling.

1.2   Semantic Interoperability

Interoperability is generally understood as being constituted by an agreement of multiple parties (platforms, code developers or similar) on a common standard, so that certain issues can be dealt with by all of them in the same way or, at least, in a sufficiently similar way. Ideally, this is the case when a whole community coherently adopts a single approach. This is often also called compatibility; in the strict sense, however, more recent use of the term compatibility restricts itself to the capability of exchanging data bilaterally, in the absence of a community standard. Theoretically, compatibility would then be more immediate than interoperability, since an intermediate third-party standard would not be required. However, it can be doubted whether this is a particularly useful distinction. Virtually every work on compatibility eventually aims at the widespread acceptance of a standard, protocol or file format. In this sense, interoperability is simply another, more modern word for all efforts at ensuring that heterogeneous software architectures, in the broadest sense, can function correctly.

Kerber and Schweitzer summarize that "interoperability has become a buzzword in European policy debates on the future of the digital economy" where "one of the difficulties of the interoperability discussion is the absence of a clear definition of interoperability" [24]. This is certainly not coincidental. A research and development landscape dominated by project-based funding from calls with priorities driven by political or cultural trends is a sure recipe for rendering the associated terminology vague to the point of complete dilution. "All stakeholders," to use another buzzword, aim at securing their share. This is evidenced by the multitude of researchers who have only recently detected that their traditional line of work is actually a subdiscipline of artificial intelligence, Industry 4.0 or data science (or, of course, all the three). In the case of interoperability, this is particularly ironic given that one of its core elements consists in defining the precise meaning of concepts. But it is not a time for academic rigour.

As understood by the present work—necessarily in disagreement with others—there are three aspects of interoperability, corresponding to the major branches of theoretical linguistics: syntax, semantics and pragmatics. Syntactic interoperability is based on a common agreement on the grammar of a formal language, such as a file format or the arrangement of data items in a stream or in memory, while semantic interoperability refers to an agreement on the meaning or implications of the communicated content. In the context of digitalization and the design of digital infrastructures, the focus is typically on establishing a shared formalization of the semantics (rather than syntax) for a particular application area, i.e. a domain of knowledge. Semantic interoperability can only be achieved if there are metadata standards by which the annotation of data is carried out and understood by all participants in an agreed way [16, 17]. This permits the integration of data communicated to a single platform from multiple sources or by multiple users. This leads to interoperability between multiple platforms whenever the developers of these platforms agree on the same metadata standards or, where semantic heterogeneity remains, if an alignment can be constructed to harmonize the divergent standards [25–29]. Accordingly, the meaning of concepts and relations needs to be agreed upon, while the technical implementation and I/O are permitted to adhere to a variety of specifications and formats.

Semantic metadata standards are also known as semantic assets; in Fig. 1.1, the most common types of semantic assets are arranged by two main measures of their expressivity and richness in content: First, the depth of the provided representation of domain knowledge; second, the depth of digitalization, characterized by the extent to which processing of the represented knowledge can be automated. At the minimum with respect to both coordinates, only a list of concepts is compiled, i.e. a vocabulary (or lexicon). If explanations and definitions are added in a way that is understandable to human readers, this becomes a dictionary; in the field of materials modelling, this includes the molecular model database (MolMod DB) nomenclature [4]. A hierarchy of concepts is a taxonomy, where multiple narrower concepts are subsumed (symbol $\sqsubseteq$) under a broader concept, yielding a tree structure, e.g. in the scientific taxonomy of biological organisms

$$\text{homo sapiens} \sqsubseteq \text{homo} \sqsubseteq \text{hominid} \sqsubseteq \text{primate} \sqsubseteq \text{mammal} \sqsubseteq \text{animal}. \tag{1.1}$$

A thesaurus extends a system of concepts by definitions of possible relations between individuals (objects) that instantiate them. For the use on digital platforms, this is typically further formalized either as a hierarchical schema or as an ontology. In a hierarchical notation (e.g. XML or JSON), relations take the form of containment, e.g. in XML format, the tag representing one object can contain tags representing subordinate objects, in an arrangement that is well defined by an XML Schema Definition (XSD) and distinct from the taxonomic hierarchy of concepts. Applied to the structure of a document, e.g. such a hierarchy might be given by

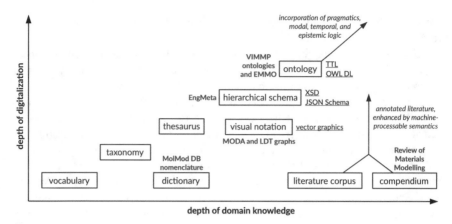

Fig. 1.1 Common categories of semantic assets. The horizontal axis indicates the amount of domain knowledge that is covered, while the vertical axis shows to what extent this information is machine-actionable. Labels adjacent to the boxes refer to formats or languages that are typically used (underlined) and to specific semantic assets that are particularly relevant to materials modelling (bold)

$$\text{word} \ \leftarrow \ \text{sentence} \ \leftarrow \ \text{paragraph} \ \leftarrow \ \text{section} \ \leftarrow \ \text{chapter} \ \leftarrow \ \text{book}, \qquad (1.2)$$

where the symbol $\leftarrow$ indicates that one entity is subordinate to another in terms of containment. In a hierarchical schema, this structural containment coexists with taxonomic subsumption. Ontologies, on the other hand, are non-hierarchical schemas, formalizing rules and definitions underlying knowledge graphs, where nodes representing individuals are connected to each other by edges that respresent the relations. The Description Logic (DL) variants that are used to specify ontologies, mainly by means of the Web Ontology Language (OWL), are more expressive than the languages used for hierarchical schemas, and can therefore, in principle, encode more domain knowledge [30, 31]. Some extensions of DL even include modal logic or temporal logic [32].

However, an agreement on syntax and semantics is insufficient without a general understanding of performative roles, the context in which a communication occurs, and generally what the different participants in an exchange can reasonably expect from each other. The statement "the accused is guilty of high treason," for instance, is syntactically correct in the English language. Its denotational meaning might be formalized by linking "the accused" to a formal representation of the specific person, and "is guilty of" and "high treason," respectively, to a relation and an entity from an ontology representing the laws of the country. However, even assuming that we accept the statement to be true, its impact will vary greatly depending on who says it (e.g. a journalist, the prosecutor or the judge), at which point, and in which context. If multiple countries decide to set up a joint court, they need to agree on the legal framework and on the language to be used at its sessions, but also on the pragmatics, much of which relates to role definitions and standards for good and best practices: How is a person appointed to become a judge, what qualifications are needed and what code of conduct needs to be followed? Pragmatic interoperability concerns such requirements and recommendations pertaining to the practice of communicating and dealing with data [33–35]. If this is to be implemented in a machine-processable way, this is inseparable from semantic technology, and closely related techniques can be used to specify semantic and pragmatic interoperability standards [35–37].

1.3 Semantic Assets and Metadata Categories

The purpose of semantic assets and metadata models, in particular, is the description of a research object in all its relevant aspects. It is advisable to define categories for this description, since the aspects might differ in their specificity. Some are general or subject to every discipline (such as file size or authorship), whereas others only apply to a single domain. Also, existing metadata standards and ontologies usually cover specific aspects of a description and then may be used as building blocks. Moreover, in big data science, automated extractability of semantic information gets crucial, and the different aspects described are differently hard to extract. In the following, we categorize the semantic description in four main classes. These originally stem from computational engineering [38], but also hold for materials modelling:

1. **Technical metadata** describe technical characteristics of the research asset, i.e. basically, the file attributes on a filesystem level and other syntactic information. These can not only be file sizes, checksum information, storage location or access dates, but also file formats.
2. **Descriptive metadata** provide general information about the research asset, such as the authors of the data, some keywords or a title. The data are described content-wise from a higher logical standpoint.
3. **Process metadata** describe the generation process and the provenance of the research asset, for example, the computational environment and software used to generate or process the data. This description may include several linked, consecutive steps.
4. **Domain-specific metadata** describe the research objects from the domain-specific perspective. In computational engineering, this includes details about the simulated target system, the simulation method or the spatial and temporal resolution, for example.

These four dimensions are the core of every rich data description. The four classes are found to hold not only for engineering but also for different fields of science. It is now subject to the metadata engineer to fill the categories with content, and we will learn how to do this in Chap. 2 taking the example of EngMeta.

The specificity of the categories is in ascending order (1–4), which is also shown in Fig. 1.2. Whereas the technical and descriptive categories and their metadata keys are generic and hold for different fields of science, category of process information is heavily bound to the research process and the domain-specific category to the research object. However, the content of the classes may overlap and a metadata

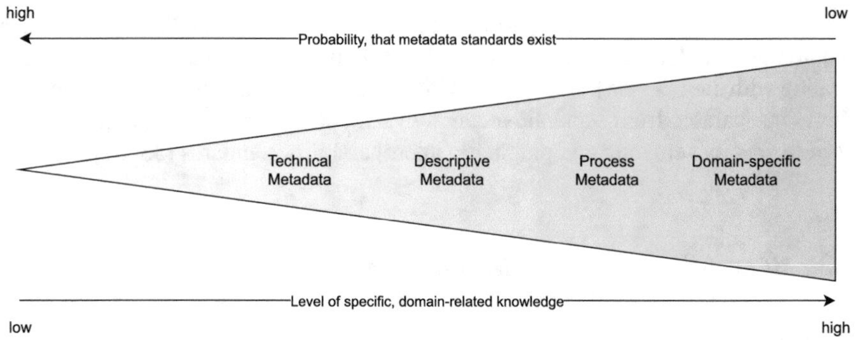

Fig. 1.2 The four metadata categories and their level of specificity

Table 1.1 Examples of relevant metadata standards for the four categories

Metadata category	Standard	Description/Remarks
Descriptive	DataCite	DOI compatibility
	DublinCore	
Technical	PREMIS	Preservation MD
Process	PROV	Provenance MD
	CodeMeta	For code and software
Domain	EngMeta	For computational engineering
	TEI	For digital humanities

key might be part of two or more categories. The probability that suitable standards exist for the four categories is decreasing with the categories' specificity, as shown in Fig. 1.2. Whereas for technical and descriptive metadata, many standards exist, this does not hold for the latter two categories, which require a significant dedicated development effort. Regarding technical metadata, the semantic information is similar in all research fields as long as the data are organized in files. A typical standard here is PREMIS [39]. Also descriptive metadata keys are similar (or even the same) throughout all disciplines. Here, DataCite is the de facto standard for a general description and citable data objects [40]. In contrast, process metadata are strongly related to the research process, where metadata standards only exit for specific processes, e.g. CodeMeta [41] and the Citation File Format [42] for the description of software and codes. For domain-specific metadata, only standards for specific research objects exist. Some relevant standards for to the four categories are shown in Table 1.1. Knowledge of the categories and existing standards enables the metadata designer to use certain parts as building blocks when compiling a standard for a certain area.

Moreover, the distinction between these four categories is crucial with respect to automated extractability. It has been shown that some categories are easier to automatically extract than others in computational engineering [38]: Technical information is

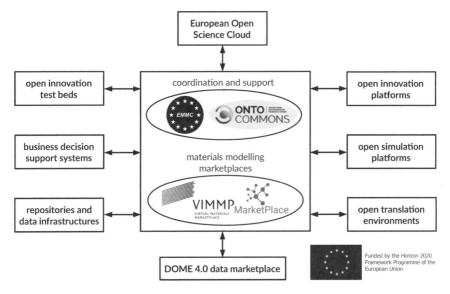

Fig. 1.3 Landscape of interoperable platforms and infrastructures in materials modelling funded from the Horizon 2020 research and innovation programme

easy to extract, since it is mostly file system attributes. Process- and domain-specific information is relatively easy to extract automatically for computational engineering applications, since it is available in output, job (input) and log files of simulation codes.[1] Descriptive information is hardly extractable, since it describes the research from a higher level and makes human interaction necessary.

1.4   Perspective and Outline of the Book

This work presents two approaches to metadata standardization in materials modelling: first, hierarchical schemas, represented here by EngMeta, an XML schema for data management in the engineering sciences that is presently in use for the DaRUS data infrastructure at the University of Stuttgart. Second, ontologies, represented by the metadata standards from the Virtual Materials Marketplace (VIMMP) project.

VIMMP belongs to the LEIT-NMBP line of the European Union's Horizon 2020 research and innovation programme, where substantial efforts and funds have been concentrated on the single objective of creating what may be labelled an environment or ecosystem of interoperable CME/ICME platforms, cf. Fig. 1.3. This line of work is grouped around two coordination and support actions (i.e. networking-oriented projects), the first of which, EMMC-CSA, led to the creation of the European Mate-

---

[1] However, this always depends on the specific simulation code and its verbosity.

rials Modelling Council (EMMC) as an interest group and community organization; the second, OntoCommons, supports the uptake of ontologies as a technology in materials modelling. The associated interoperability effort is based on the Review of Materials Modelling (RoMM), a compendium that aims at establishing a coherent understanding of all major modelling and simulation approaches from quantum mechanics up to continuum methods [43]. On this basis, first, MODA (Model Data) was introduced as a standardized description simulation workflows together with their intended use cases [44]; subsequently, a variety of domain ontologies were developed and connected to the European Materials and Modelling Ontology (EMMO), a top-level ontology aiming at describing *all that exists* from a perspective that is advantageous for CME/ICME infrastructures and applications [45, 46]. EngMeta, on the other hand, was developed at the University of Stuttgart in an environment where digitalization of materials modelling is advanced through the Cluster of Excellence "Data-Integrated Simulation Science" (EXC 2075), the Stuttgart Center for Simulation Science, and work on repositories including ReSUS (Reusable Software University Stuttgart), DaRUS (Data Repository of the University of Stuttgart), and the programme for national research data infrastructures (NFDI) of the German Research Foundation (DFG).

Both approaches have their strengths. On the one hand, ontologies are experiencing a great surge in popularity. They are increasingly seen as a key component of state-of-the-art solutions in data technology. Nonetheless, they also have drawbacks. First, as a technology, ontologies are comparably heavy, requiring substantial resources for development and maintenance. While certain communities have succeeded at establishing agreed domain-specific semantic frameworks [47, 48], it is also commonly found that (occasionally quite complex) ontologies are developed within a project and then abandoned when the project is over. While it is undeniable that classification schemes, in general, are a prerequisite for interoperability, there is no consensus on whether a less expressive framework or a more expressive one is to be preferred. The advantage of less expressive languages is they can be handled with multiple technologies and tools, and are typically lighter and faster. Richer languages allow to describe more complex relations, but at the price of being tied to newer, less widespread technologies and being typically computationally more demanding. Moreover, all new technologies need to overcome a barrier to adoption before they are employed widely. Superficially, it may seem that ontologies have advanced relatively far on this path; however, comparing the uptake of ontology-based semantic technologies in research software engineering (simulation codes, etc.) with alternatives such as XML/XSD-based solutions makes it clear that the advance of ontology-based solutions is still at an early stage.

Additionally, ontology design can result in an overregulation of domain practices. This risk is inherent to the prescriptive (rather than descriptive) nature of ontologies and taxonomies—like grammar regulates syntax, ontologies pretend to regulate meaning. In reality, however, there are always many possible ways to ontologize any given domain of knowledge. The concurrent development of multiple incommensurable paradigms is one of the manifestations of progress in a scientific discipline and the major driving force for the emergence of new specializations [49]; in the words of Kuhn, scientific revolutions are characterized by a "change in several of the taxonomic categories prerequisite to scientific descriptions and generalizations. That change, furthermore, is an adjustment not only of criteria relevant to categorization, but also of the way in which given objects and situations are distributed among pre-existing categories" [50]. Insisting on the adoption of a single ontology by a whole field of science makes that field perfectly interoperable, but at the price of scientific stagnation. In this view, any field that is developing fast—as modelling and simulation technology today certainly is—will consider a plurality of paradigms, *semantic heterogeneity*, as an indicator of its success, rather than as an obstacle, and pursue the frequent renegotiation of alignments between multiple ontologies, taxonomies and hierarchical schemas [26, 27, 51].

However, even if knowledge is formalized in a machine-readable format, besides the design of the ontology, there will still be other steps where human intervention is needed, notably, to classify and annotate the individuals. Whenever this occurs, we rely on the fact that the semantics is really shared, i.e. that there is an actual common understanding of all the concepts by all the users. In practice, colloquial language and insufficient familiarity with concept definitions can give rise to common pitfalls and false friends in using ontologies. By stating that something is *generally* true, we could mean "all the time" or "most of the time;" in the EMMC context, the words *translation* and *mesoscopic* have very specific meanings that would be misconstrued intuitively by most domain experts in materials modelling. Hence, human error must be anticipated (in addition to genuine disagreements on how an ontology should be applied), and similarly, automated annotation tools based on natural language processing are not free of error. All of this highlights the need for a community to gather and share concepts in a continuous effort; it corroborates the conclusion that there is a trade-off between the expressive power of ontologies and the associated social and technological cost of development and maintenance.

The remainder of the book is structured as follows: Chapter 2 introduces the XSD-based engineering metadata schema EngMeta and the research data infrastructure DaRUS, situating it in the context of the emerging environment of databases and repositories for data from physics-based modelling and simulation. The system of marketplace-level domain ontologies developed by VIMMP is presented in Chap. 3, concerning data provenance, services and transactions at the marketplace level, and in Chap. 4, concerning the description of solvers, associated aspects such as licenses and software features, and the characterization of physical and other variables that occur in modelling and simulation. On this basis, Chap. 5 addresses issues related to the practical use of the metadata standards, including syntactic interoperability and concrete scenarios from molecular modelling and simulation; it also discusses challenges that arise from semantic heterogeneity, wherever multiple interoperability standards are concurrently employed for identical or overlapping domains of knowledge, or where domain ontologies need to be matched to top-level ontologies such as the EMMO.

References

1. E. Forte, F. Jirasek, M. Bortz, J. Burger, J. Vrabec, H. Hasse, Digitalization in thermodynamics. Chem. Ing. Tech. **91**(3), 201–214 (2019)
2. N. Tchipev, S. Seckler, M. Heinen, J. Vrabec, F. Gratl, M. Horsch, M. Bernreuther, C.W. Glass, C. Niethammer, N. Hammer, B. Krischok, M. Resch, D. Kranzlmüller, H. Hasse, H.J. Bungartz, P. Neumann, TweTriS: twenty trillion-atom simulation. Int. J. HPC Appl. **33**(5), 838–854 (2019)
3. Å. Ervik, A. Mejía, E.A. Müller, Bottled SAFT: a web app providing SAFT-γ Mie force field parameters for thousands of molecular fluids. J. Chem. Inf. Model. **56**(9), 1609–1614 (2016)
4. S. Stephan, M.T. Horsch, J. Vrabec, H. Hasse, MolMod - an open access database of force fields for molecular simulations of fluids. Mol. Sim. **45**(10), 806–814 (2019)
5. A. Merkys, N. Mounet, A. Cepellotti, N. Marzari, S. Gražulis, G. Pizzi, A posteriori metadata from automated provenance tracking: integration of AiiDA and TCOD. J. Cheminform. **9**, 56 (2017)
6. M.T. Horsch, C. Niethammer, G. Boccardo, P. Carbone, S. Chiacchiera, M. Chiricotto, J.D. Elliott, V. Lobaskin, P. Neumann, P. Schiffels, M.A. Seaton, I.T. Todorov, J. Vrabec, W.L. Cavalcanti, Semantic interoperability and characterization of data provenance in computational molecular engineering. J. Chem. Eng. Data **65**(3), 1313–1329 (2020)
7. X. Xu, J. Range, G. Gygli, J. Pleiss, Analysis of thermophysical properties of deep eutectic solvents by data integration. J. Chem. Eng. Data **65**(3), 1172–1179 (2019)
8. G.J. Schmitz, Microstructure modeling in integrated computational materials engineering (ICME) settings: can HDF5 provide the basis for an emerging standard for describing microstructures? JOM **68**(1), 77–83 (2016)
9. G.J. Schmitz, B. Böttger, M. Apel, J. Eiken, G. Laschet, R. Altenfeld, R. Berger, G. Boussinot, A. Viardin, Towards a metadata scheme for the description of materials: the description of microstructures. Sci. Technol. Adv. Mater. **17**(1), 410–430 (2016)
10. J. Morbach, A. Wiesner, W. Marquardt, OntoCAPE 2.0: a (re−)usable ontology for computer-aided process engineering. Comput. Aid. Chem. Eng. **25**, 991–996 (2008)
11. J.P. Belaud, M. Pons, CAPE-OPEN: interoperability in industrial flowsheet simulation software. Chem. Ing. Tech. **86**(7), 1052–1064 (2014)
12. L. Koo, N. Trokanas, F. Cecelja, A semantic framework for enabling model integration for biorefining. Comput. Chem. Eng. **100**, 219–231 (2017)
13. P.B. Heidorn, Shedding light on the dark data in the long tail of science. Libr. Trends **57**(2), 280–299 (2008)
14. B. Schembera, J.M. Durán, Dark data as the new challenge for big data science and the introduction of the scientific data officer. Philos. Technol. **33**, 93–115 (2020)
15. J. Bicarregui, Building and sustaining data infrastructures: putting policy into practice. Policy document (2016). https://doi.org/10.6084/m9.figshare.4055538.v2
16. B. Mons, *Data Stewardship for Open Science* (CRC Press, Boca Raton, 2018)
17. G. Guizzardi, Ontology, ontologies and the "I" of FAIR. Data Intell. **2**(1–2), 181–191 (2020)
18. B. Bouchou, C. Niang, Semantic mediator querying, in *Proceedings of IDEAS '14*, ed. by A.M. Almeida, J. Bernardino, E. Ferreira Gomes (ACM, New York, USA, 2014), pp. 29–38
19. G. Fusco, L. Aversano, An approach for semantic integration of heterogeneous data sources. Peer J. Comput. Sci. **6**, e254 (2020). https://doi.org/10.7717/peerj-cs.254
20. D. Lembo, R. Rosati, V. Santarelli, D.F. Savo, E. Thorstensen, Mapping repair in ontology-based data access evolving systems, in *Proceedings of IJCAI*, IJCAI, ed. by C. Sierra (San José, California, USA, 2017), pp. 1160–1166
21. G. Xiao, D. Calvanese, R. Kontchakov, D. Lembo, A. Poggi, R. Rosati, M. Zakharyaschev, Ontology-based data access: a survey, in *Proceedings of IJCAI*, IJCAI, ed. by J. Lang (San José, California, USA, 2018), pp. 5511–5519
22. A. Ribes, C. Caremoli, Salomé platform component model for numerical simulation, in *Proceedings of COMPSAC 2007*, IEEE Computer Society, vol. 2, ed. by C.K. Chang (Los Alamitos, California, USA, 2007), pp. 553–564

23. G. Pizzi, A. Cepellotti, R. Sabatini, N. Marzari, B. Kozinsky, AiiDA: automated interactive infrastructure and database for computational science. Comput. Math. Sci. **111**, 218–230 (2016)
24. W. Kerber, H. Schweitzer, Interoperability in the digital economy. Jipitec **8**(1), 39–58 (2017)
25. F.M. Suchanek, S. Abiteboul, P. Senellart, PARIS: probabilistic alignment of relations, instances, and schema. Proc. VLDB Endow. **5**(3), 157–168 (2011)
26. J. Euzenat, P. Shvaiko, *Ontology Matching*, 2nd edn. (Springer, Heidelberg, 2013)
27. F. Duchateau, Z. Bellahsene, YAM: a step forward for generating a dedicated schema matcher, in *Transactions on Large-Scale Data- and Knowledge-Centered Systems XXV*, ed. by A. Hameurlain, J. Küng, R. Wagner, LNCS, vol. 9620 (Springer, Heidelberg, Germany, 2016), pp. 150–183
28. M. Koutraki, N. Preda, D. Vodislav, Online relation alignment for linked datasets, in *Proceedings of ESWC 2017*, ed. by E. Blomqvist, D. Maynard, A. Gangemi, R. Hoekstra, P. Hitzler, O. Hartig, LNCS, vol. 10249 (Springer, Cham, Switzerland), pp. 152–168 (2017)
29. L. Zhou, M. Cheatham, P. Hitzler, Towards association rule-based complex ontology alignment, in *Proceedings JIST 2019*, ed. by X. Wang, F.A. Lisi, G. Xiao, E. Botoeva, LNCS, vol. 12032 (Springer, Cham, Switzerland, 2020), pp. 287–303
30. D. Allemang, J. Hendler, *Semantic Web for the Working Ontologist*, 2nd edn. (Morgan Kaufmann, Waltham, Massachusetts, USA, 2011)
31. F. Baader, I. Horrocks, C. Lutz, U. Sattler, *An Introduction to Description Logic* (Cambridge University Press, Cambridge, 2017)
32. T. Schneider, M. Šimkus, Ontologies and data management: a brief survey. Künstl. Intell. **34**(3), 329–353 (2020). https://doi.org/10.1007/s13218-020-00686-3
33. M. Schoop, A. de Moor, J. Dietz, The pragmatic web: a manifesto. Commun. ACM **49**(5), 75–76 (2006)
34. F. Weidt Neiva, J.M.N. David, R. Braga, M.R.S. Borges, F. Campos, SM2PIA: a model to support the development of pragmatic interoperability requirements, in *Proceedings of ICGSE 2016*, ed. by D. Redmiles, M.A. Gerosa, T. Hildenbrand (IEEE, New York, USA), pp. 119–128
35. F. Weidt Neiva, J.M.N. David, R. Braga, F. Campos, Towards pragmatic interoperability to support collaboration: a systematic review and mapping of the literature. Inf. Softw. Technol. **72**, 137–150 (2016)
36. P. De Leenheer, S. Christiaens, Mind the gap! Transcending the tunnel view on ontology engineering, in *Proceedings of ICPW '07*, ed. by S. Buckingham Shum, M. Lind, H. Weigand (ACM, New York, USA, 2007), pp. 75–82
37. M. Gan, Enterprise isomorphic mapping mechanism: towards ontology interoperability in EIS development, in *Proceedings of ICEBE*, IEEE Computer Society, ed. by P. Kellenberger (Los Alamitos, California, USA, 2009), pp. 340–345
38. B. Schembera, Forschungsdatenmanagement im Kontext dunkler Daten in den Simulationswissenschaften. Dissertation, Universität Stuttgart (2019). https://doi.org/10.18419/opus-11028
39. B. Lavoie, R. Gartner, *Preservation Metadata*, DPC Technology Watch Series, Digital Preservation Coalition, 2nd edn. (NewYork, UK, 2013)
40. J. Neumann, J. Brase, DataCite. Names for research data. J. Comp.-Aid. Mol. Des. **28**(10), 1035–1041 (2014)
41. M.B. Jones, C. Boettiger, A. Cabunoc Mayes, A. Smith, P. Slaughter, K. Niemeyer, Y. Gil, M. Fenner, K. Nowak, M. Hahnel, L. Coy, A. Allen, M. Crosas, A. Sands, N.C. Hong, P. Cruse, D.S. Katz, C. Goble, CodeMeta: an exchange schema for software metadata. Version 2.0. Technical report, KNB Data Repository (2017). https://doi.org/10.5063/schema/codemeta-2.0
42. S. Druskat, N.C. Hong, R. Haines, J. Baker, Citation file format (CFF): specifications. Technical report, Zenodo (2018). https://doi.org/10.5281/zenodo.1405679
43. A.F. De Baas (ed.), *What makes a material function?* (EU Publications Office, Luxembourg, Let me compute the ways, 2017)
44. CEN-CENELEC Management Centre, Materials modelling: terminology, classification and metadata. CEN workshop agreement 17284, Brussels, Belgium (2018)
45. G. Goldbeck, E. Ghedini, A. Hashibon, G.J. Schmitz, J. Friis, A reference language and ontology for materials modelling and interoperability, in *Proceedings of NWC 2019*, NAFEMS (Knutsford, UK, 2019), p NWC_19_86

46. EMMC Coordination and Support Action, European Materials and Modelling Ontology (2020), https://github.com/emmo-repo/ and https://emmc.info/emmo-info/. Accessed 8 Apr 2020

47. OBO Technical WG, Open Biological and Biomedical Ontology (OBO) Foundry (2020). http://www.obofoundry.org/. Accessed 23 Mar 2020

48. B. Smith, M. Ashburner, C. Rosse, J. Bard, W. Bug, W. Ceusters, L.J. Goldberg, K. Eilbeck, A. Ireland, C.J. Mungall, N. Leontis, P. Rocca-Serra, A. Ruttenberg, S.A. Sansone, R.H. Scheuermann, N. Shah, P.L. Whetzel, S. Lewis, The OBO Foundry: coordinated evolution of ontologies to support biomedical data integration. Nat. Biotechnol. **25**(11), 1251–1255 (2007)

49. A. Davies, Kuhn on incommensurability and theory choice. Stud. Hist. Philos. Sci. **44**(4), 571–579 (2013)

50. T.S. Kuhn, What are scientific revolutions? in *The Probabilistic Revolution*, ed. by L. Krüger, L. Daston, M. Heidelberger (MIT Press, Cambridge, Massachusetts, Germany, 1987), pp. 7–12, 19–21

51. N.F. Noy, Semantic integration: a survey of ontology-based approaches. SIGMOD Rec. **33**(4), 65–70 (2004)

Chapter 2
Research Data Infrastructures and Engineering Metadata

The two core elements of data technology in every field of science, in general, and in materials modelling, in particular, are metadata or ontologies and data infrastructures. Even though they can work independently, they are strongly connected. Whereas metadata describes the data, the task of the research data infrastructure is to store and to preserve the data and to connect it with its metadata description. So, mere data becomes semantically interoperable and therefore a valuable piece of information respecting the FAIR principles.

The chapter introduces metadata models as a semantic technology for knowledge representation to describe selected aspects of a research asset in Sect. 2.1. The process of building a hierarchical metadata model is re-enacted in this chapter and highlighted by the example of EngMeta [1]. Moreover, this chapter gives an overview on data infrastructures in Sect. 2.2. In this section, the general architecture and functions are discussed and multiple examples of data infrastructures in materials modelling are given.

2.1 Engineering Metadata

This section examines engineering metadata. The term is ambiguous on purpose. First, engineering metadata names metadata which is used for engineering applications, such as materials modelling. Second, engineering metadata conceptualizes the art of designing metadata in a more general way.

This section is organized as follows. First, it is described how an ontology-based metadata model is created in a general way in Sect. 2.1.1. Second, this process is explained along EngMeta, a metadata model for engineering in Sect. 2.1.2.

© The Author(s) 2021
M. Horsch et al., *Data Technology in Materials Modelling*,
SpringerBriefs in Applied Sciences and Technology,
https://doi.org/10.1007/978-3-030-68597-3_2

2.1.1   How to Engineer Metadata

The art of engineering a metadata model includes several consecutive steps which are described in this subsection. It may happen that this process or a single step has to be iterated several times to come to a fine-grained, purposeful description of the research asset. In short, the following steps are necessary to engineer a metadata model. First, a consensus must be reached about what metadata actually serves in the single context. Then, an object model has to be carved out of the research process. Last, the object model has to be transferred to a formal representation and implemented and therefore becomes a metadata model.

2.1.1.1   Definitions of Metadata and Metadata Models

However, in the beginning of designing metadata for a certain purpose, it first has to be discussed how metadata is defined. Usually, metadata is defined as a structured form of knowledge representation, or simply, as many authors put it, "data about data" [2]. Edwards describes this as the holy grail of information science:

> Extensive, highly structured metadata often are seen as a holy grail, a magic chalice both necessary and sufficient to render sharing and reusing data seamless, perhaps even automatic. [3, p. 672]

However, metadata is always strongly context dependent. To tackle their context dependence, metadata must serve as a mode of communication:

> We propose an alternative view of metadata, focusing on its role in an ephemeral process of scientific communication, rather than as an enduring outcome or product. [3, p. 667]

Following this, metadata takes the role of semantic technology: Its task is to relieve the direct communication and negotiation of data producers and data consumers and should therefore diminish "science friction" [3], which occurs in every process where research data is exchanged. To illustrate science friction, imagine two researchers exchanging a dataset, which is not properly described by metadata. The receiver might suppose the variable t_i as a data point in a time series. To provide clarification, the receiver would have to contact the sender of the data, and also in this process can be defective. This example shows the importance of metadata as semantic asset, and therefore as a mode of fixed, negotiated communication.

Additionally, as Jane Greenberg puts it, metadata should semantically support the specific workflow [4]. For example, metadata describes a data point with an error bar and defines the form of the error. Thus, metadata would support the interpretation of the data point.

Following the discussion of metadata, a metadata model then can be seen as the middle ground of a non-formal model and a complete formalization of metadata keys, according to [5]. Its task is to describe the research objects or parts of it and its relation to other objects. They are still interpretations; however, they are constructed in a transparent and comprehensible way and derived from a common understanding of the research object, and lead to a fixed negotiation. The approach described in this chapter could also be called an ontology-based metadata, since the metadata model is engineered from an object model. As depicted in Fig. 1.1, hierarchical models such as EngMeta range below an ontology; however, their task is also to balance the depth of domain knowledge representation and the depth of digitization. The question in what terms a metadata model is different from an ontology has already been discussed in Sect. 1.2.

2.1.1.2 Object Model

The object model is the starting point for engineering a metadata model and marks the first phase in the creation process [5]. In this phase, an object model, respectively, an ontology description is carved out in a non-formal or natural language (and maybe containing graphical elements) describing and explicating all the relevant objects, terms, relations and rules. Every person potentially involved has to contribute to this process, since the metadata model will act as a semantic convention for a common understanding of the research data described.

The first part of engineering an object model is a clear and fixed understanding of what the object of research is, and what data it is representing. This can only be conducted by the analysis of the research process with all the stakeholders included. In this step, following information must be gathered:

- **Entities** All relevant entities (or objects) of the research process must be identified. This includes finding classes of entities, grouping entities or merging them. In materials modelling, one entity which is relevant is, for example, the *component* which represents a chemical species.
- **Attributes** For each entity defined in the previous step, attributes describing the entity must be found. To stick with the example, the *component* is characterized by attributes like a *name*, the *smiles* or *IUPAC* code and a *unit*.
- **Relations** In this part, the relations between the entities must be cleared, e.g. how they are linked to each other to deliver a holistic description. The arguments must be reasonable, but are strongly specific to the research. For example, one could argue that the *component* is related to the *simulated target system*. Usually in metadata modelling, *is-part-of* relations are sufficient to model the vast majority of cases. However, relations are not limited to these hierarchical types and may give a semantically more advanced description which will eventually lead towards ontologies.

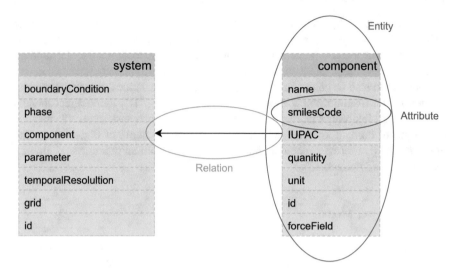

Fig. 2.1 Example of *component* entity, which has several attributes, such as the *smilesCode* and is a part of the simulated target *system*, which is shown by a relation

Figure 2.1 shows how a *component* in materials modelling could be represented by an entity, some attributes and a relation according to the example given above. All the entities can then be categorized according to the proposed classes of Sect. 1.3. The *component* entity would be categorized as discipline-specific metadata.

Also in this step, the question arises if the description needs to be data centric or process centric. It strongly depends on the research process how to answer this question. For example, in code development, one needs to continuously follow the changes made to the codes, i.e. the process of programming. Hence, the appropriate description of programming can only be process centric.[1] In data science applications, it is strongly dependent on the workflow, if a data-centric or a process-centric description should be chosen. In general, if data is the main outcome, even in a chain of process steps, one might want to choose a data-centric approach. If the processes are central to the research endeavour, and each process has a discrete output, one might chose a process-centric description. Of course, both approaches are not mutually exclusive. A data-centric approach also includes process information and a process-centric approach an elaborated description of the data. It is just a matter of hierarchical structuring and precedence. In Sect. 2.1.2, we will discuss why and how we decide for a data-centric model for computational engineering and realized in EngMeta.

---

[1] This is reflected by tools such as git, which include metadata for every commit to describe the process.

2.1.1.3   The Metadata Model and Its Implementation

When the object model is converted to a formal language, special care has to be taken if parts of the object model already exist in some standard. With respect to the categorization taken in Sect. 1.3, the probability to find existing, fitting standards for technical or descriptive metadata is high, whereas for process- and domain-specific standards they are not likely to be found. Some of the relevant standards are described in quoted section; however, an excessive amount of standards exist.

Another consideration when implementing the model is choosing the right formal language for representing the metadata model. Most likely, this will be XSD[2] or JSON Schema.[3] Both offer a strict structural definition of the entities, attributes and relations, and the decision is more or less based on setting of the metadata model: What are the skills available, what are the technical requirements for the implementation? For example, the question, which standard the database or repository supports, where the metadata later will stored, is crucial in deciding for an implementation language.

2.1.1.4   Metadata Processes

A metadata model alone is not sufficient. As Edwards puts it, metadata products such as models have to be accomplished by metadata processes:

> Metadata products can be powerful resources, but very often—perhaps even usually—they work only when metadata processes are also available. [3, p. 668]

Otherwise, if processes are not available, something called "metadata friction" would occur and the semantic assets would become worthless. This phenomenon would indicate the additional effort of (manual) metadata annotation and management, which has to be reduced by corresponding processes. This view is backed by the FAIR principles [6] and the additional guidance from an EU report [7]. The FAIR principles state metadata description as the main concept, and the study [7] accomplished this rather technical approach by processes surrounding these principles. In the case of materials modelling and computational engineering, in general, these processes would include, but are not limited to, the following:

- **Automated metadata extraction**. One finding of [8] states that manual metadata annotation is a barrier for good research data management especially in the engineering science. Hence, automated metadata extraction is a major supporting process.
- **Data and metadata stewardship**. Data and metadata need clear responsibilities and roles that define stewardship. This means that such a role has the responsibility of supporting metadata annotation, building metadata models and checking the

---

[2]https://www.w3.org/TR/xmlschema11-1/.

[3]https://json-schema.org/.

data inventory for unindexed data. Such a role is, for example, the Scientific Data Officer [9].

- **Incentives**. On main process to support metadata products is incentives to use models and tag the data with metadata. These incentives can either be intrinsic or extrinsic. Intrinsic incentives would include low barriers for metadata annotation. Extrinsic incentives would include making metadata annotation of the published research data mandatory for scientific publication.
- **Culture**. Supporting metadata annotation and also cultural processes have to be adapted. Metadata annotation and research data management have to be seen as one essential part of scientific practice. The process of science has to be adapted to 1. publishing the data Open Access and 2. applying FAIR paradigm of data description to it. However, this cultural change may be linked to the above process of incentives. As of now, researchers only get recognition for publishing papers and not the data.

2.1.2   Metadata for Engineering: The EngMeta Metadata Scheme

In this subsection, an example for a metadata model and its design will be given. EngMeta [1, 8, 10] is a semantic metadata standard for computational engineering and was designed following principles of the above subsection. Following Staab et al. [5] EngMeta could be referred to as an ontology-based metadata model. A comparison to VIMMP as a genuine ontology is carried out in Sect. 4.5. EngMeta was designed as a joint effort of researchers from computational engineering sciences (process engineering and aerodynamics), from the library sciences as well as from the computer sciences. This allowed the design of an integrated metadata model covering all the relevant research aspects in all the four categories as described in Sect. 1.3.

2.1.2.1   The Object Model of EngMeta

For the design of EngMeta, the object of research had to be identified first. This seems to be an easy task, but the devil is in the detail.

As aerodynamics and molecular dynamics served as use cases, it was clear that computational engineering and its outcome were the common ground, but not more. All the four metadata categories defined in Sect. 1.3 had to be written out with representations, which could only be accomplished by analysing the research itself for common entities and attributes for process and domain. Both technical and descriptive metadata keys were quite straightforward, since their specificity is low (see Fig. 1.2). The process metadata and the domain-specific metadata were harder to carve out from both use cases and could only be gathered by a detailed analysis of the research process. The following entities were determined as process metadata for computational engineering:

- **processingStep** serves as the highest level of the description for the provenance of the data and describes one processing step in the research process.
- **environment** describes the computational environment on which the research was conducted, e.g. the hardware and compiler.
- **software** describes the software environment in which the research was conducted, e.g. the code and its version.

The following entities were determined as domain-specific metadata for computational engineering applications and were seen as common ground, stemming from the use case of aerodynamics and thermodynamics but could also be applicable for use cases of materials modelling and beyond:

- **system** This key represents the simulated target system (or the observed system) and its characteristics, which are the metadata keys listed below.
- **variable** This metadata field represents the used variables and parameters, which can be either controlled or measured variables. This is not bound to a specific field of research but holds more generally for most applications in computational science, as variables and parameters are the basis of every simulation.
- **method** This field holds the information on the simulation method, such as "simulation with umbrella sampling".
- **component** This metadata key describes the names and SMILES/IUPAC codes of the molecules and solvents used within the simulation.
- **force field** Describes the force field which is used for the simulation.
- **boundaryCondition** Describes the properties on the boundaries of two components.
- **spacial resolution** This key defines the spacial resolution of a simulation.
- **temporal resolution** This key defines the temporal resolution of a simulation, for example, the number of timesteps, the interval and other characteristics.

It also became clear that the model will be data centric, since the research process in computational engineering reaches a steady state when a dataset is produced by a simulation or by post-processing of some data. However, it is crucial to document the processing steps as well for a good provenance description. This leads to a object model where the dataset is on top of the hierarchy and can include several processing steps.

The complete object model of EngMeta, with all entities, their attributes and relations, is depicted in Fig. 2.2. The four metadata categories are coloured differently.

2.1.2.2   The Metadata Model of EngMeta and its Implementation

After setting up the object model, research was conducted if there are metadata standards that serve the purpose of describing research assets in computational engineering as defined by the object model. None was found, however it was identified

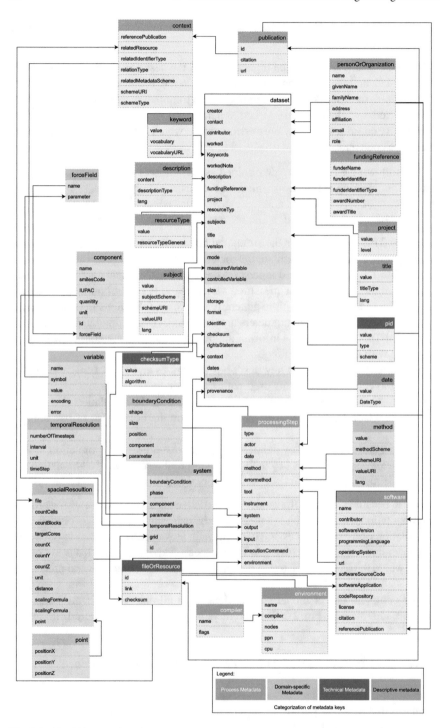

Fig. 2.2 The Object Model of EngMeta [8]

Table 2.1 Existing standards that were used in EngMeta and VIMMP with respect to the four categories defined in Sect. 1.3

	EngMeta Metadata Model	VIMMP Ontology
Technical	PREMIS	–
Descriptive	DataCite	MMTO, OTRAS, VICO
Process	CodeMeta, ExptML, UnitsML	VISO
Domain specific	–	VISO, VOV

that different metadata standards cover certain aspects of the EngMeta entities. This coverage is shown in Table 2.1 with respect to the four metadata categories. CodeMeta is a description of software tools and serves for the software part in EngMeta. Data-Cite is the standard for descriptive metadata and moreover, enables the data to get a DOI and was therefor integrated into EngMeta. PREMIS is a standard for technical metadata, and ExptML was integrated for experimental device, which can also be modelled by EngMeta. As Prov is a standard for provenance, a crosswalk for this standard was developed in order to achieve semantic interoperability [1]. Moreover, in this table, a comparison to VIMMP, which is discussed in Chap. 4 regarding existing standards is shown. The model has been implemented as an XML Schema Definition (XSD) and is available for open use and modification.[4]

2.1.2.3 The Metadata Processes Supporting EngMeta

As discussed in Sect. 2.1.1.4, a metadata model needs to be complemented with metadata processes. Otherwise, it will not be fully effective to make research data FAIR. In the example of EngMeta, the model was complemented by an automated metadata extraction, the establishment of a research data management competence centre and an institutional repository. Details on the repository can be found in the following section on research data infrastructures, especially in Sect. 2.2.3.1. FOKUS was established as the main competence centre for questions and support regarding research data management at the University of Stuttgart. The automated metadata extraction ExtractIng was designed and implemented. It works in a way that all the existing metadata, stemming from log-, job- and various other files in the HPC and simulation environment, are extracted and are converted to the EngMeta metadata model. It can be integrated in the specific research process, and it was shown how an automated approach would look like for simulation sciences. Right after the simulation run, the ExtractIng tool will be triggered, transforming all the scattered metadata in a standardized form according to EngMeta. Then, the metadata can be automatically uploaded to the repository, all together with the data, forming a dataset within the repository including all relevant semantic information for FAIR interoperability.

---

[4]https://www.izus.uni-stuttgart.de/fokus/engmeta/.

2.2   Research Data Infrastructures

Research data infrastructures enable the data to become findable and accessible (the FA in FAIR), whereas semantic standards enable the interoperability and reusability (the IR in FAIR). Hence, research data infrastructures are the second crucial pillar for FAIR data technology as both parts are inseparable for semantic interoperability in materials modelling. Research data infrastructures resemble to repositories as they ensure enriching data with metadata, long-term preservation and open-access availability for the scientific community. Moreover, the data infrastructures serve as the link between the data and the community, and therefore play a significant role in science.

This section is organized as follows. First, the requirements and functions for data infrastructures are explained in detail in Sect. 2.2.1. Then, generic architectural key characteristics are discussed in Sect. 2.2.2. Moreover, examples of research data infrastructures relevant for materials modelling are highlighted in Sect. 2.2.3.

2.2.1   Requirements and Functions

Data infrastructures in materials modelling should, besides the typical data management tasks of storing, sharing and enabling FAIR data, support the specific research by integrating open simulation codes, analytics tools and the management of the scientific workflow [11]. This means that a data infrastructure goes beyond mere archival repositories. However, the core of all data infrastructures is an archive with repositoral functions. The OAIS Reference Model (ISO 14721) can give an orientation how such a core may look like [12], and the following functionality was derived from this framework:

- **Data Ingest** Functionalities how to ingest data have to be defined and implemented. This includes the design of an appropriate user interface and integration in the workflow.
- **Data Preservation and Archiving** Originally split into two functionalities in the OAIS framework, for our purpose of defining functionalities for materials modelling, merging them into one is sufficient. This functionality should ensure permanent storage of the ingested data. Data preservation resembles to bitstream preservation on this layer.
- **Data Management** This functionality corresponds to metadata management and linking the data objects according to metadata information.
- **Administration** This functionality includes not only administrative tasks, but also policy management and AAI.

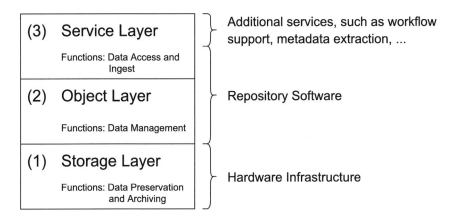

Fig. 2.3 Layers, functions and technologies for data infrastructures. The three layers (storage, object and service) are related to their functions defined in Sect. 2.2.1 and respective technological coverage of hardware, software and services

- **Data Access** This functionality must be designed and implemented by a user interface in order to ensure data access for users. Moreover, this includes capabilities to search and explore the data infrastructure.

As it was mentioned earlier, the above basic functions have to be accompanied by supportive functions for the scientific workflow. These should include the following:

- **Workflow support** This means that the above functionalities have to be integrated seamlessly into the scientific workflow in the field.
- **Service tool integration** As moving data is expensive, the data infrastructure has to enable data analytics and processing tools close to the data repository. This can also include visualization services.

2.2.2 Architectures

Data infrastructures can be logically divided into three major layers, which are depicted in Fig. 2.3 [13]. The functions defined in the previous Sect. 2.2.1 have to be implemented in the specific or throughout all the three layers. It is subject to the precise implementation of a data infrastructure which function resides in which layer.[5]

The base layer of a data infrastructure is the storage layer (l1), where the data objects are physically stored and bitstream preservation is guaranteed. Technically, this layer can exist in distributed and/or hierarchical setting and is often a combination from hard disc and tape storage. The intermediate layer is the object layer (l2), whose

---

[5]Mapping these functions to layers is not trivial, an example can be found in [14].

Table 2.2 Data repository software

Repository	Origin	Sample installation (Type[,field])
Dataverse	Data management	University of Stuttgart/DaRUS (institutional)
Dspace	Document management	Fraunhofer Gesellschaft/Fordatis (institutional)
Fedora	Document management	Saarland University/CLARIN (domain-specific, linguistics)
Invenio	Data management	Swiss National Computing Centre/Materials Cloud ARCHIVE (domain specific, materials modelling)

basic functionality is metadata management. By this layer, data from the storage layer is enriched with metadata and data objects become information objects with a persistent identifier, whose purpose is to make the data citable. The third layer is the service layer (l3) and includes the user interface and marks the visible part of the data infrastructures. Moreover, this layer includes additional services, such as an automated metadata extraction.

Basically, data infrastructures implement all three layers; however, they can operate or work in distributed environments. Usually the base layer (l1) is the hardware part of the data infrastructure, whereas the layers (l2) and (l3) are the software part. The functionalities of the layers (l2) and (l3) are usually covered by repository software. A repository is a store for data that organizes this data in some logical manner and makes the data available for usage to a specified group of persons. It is important to mention that a repository is *not* a filesystem, which means that its purpose is not to manage the files in directory structures. In contrast, a repository must be imagined as collections of files organized in sets (of some logical manner, for example, as datasets, as linked data, in a loose hierarchical structure,...), which are described by metadata, are search and retrievable, and are provided with a persistent identifier.

Out-of-the-box generic repository software packages are generally available and serve different purposes. Some of those packages stem from document management, whereas others have their origins in data/file management. Their origin has to be taken into account when evaluating the repository for a specific use case or domain. Table 2.2 gives an overview over typical data repository software. For example, Dataverse originates from the management of datasets, whereas Dspace stems from managing document files.[6] However, also Dspace is capable of managing datasets, and the

---

[6]In the context of this chapter, iRods has to be mentioned. Even though it is not a classical repository software package but offers a unified namespace, its functionalities include repository-style data management on a filesystem level [15].

Fraunhofer-Gesellschaft is using it to store research data in its institutional repository *Fordatis*[7] [16].

Research data infrastructures can be classified as institutional and domain-specific infrastructures. Institutional data infrastructures resemble to research data management on an institutional level and are not bound to a specific discipline. An example of this type is DaRUS, which will be discussed in Sect. 2.2.3.1 of this chapter. A domain-specific data infrastructure serves as an approach which is bound to a specific discipline and can span across multiple institutions. An example of a domain-specific data infrastructure for materials modelling is NOMAD, which will be discussed in Sect. 2.2.3.2.

2.2.3 Examples of Research Data Infrastructures in Materials Modelling

2.2.3.1 DaRUS

Even though the *Data Repository of the University of Stuttgart (DaRUS)*[8] is an institutional repository and not limited to materials modelling, it will be discussed here since its development was strongly driven by the EngMeta metadata model. Moreover, it is an example of a loosely coupled data infrastructure. Its overall development was urged by the need of a sustainable repository for the University of Stuttgart and, in particular, the materials modelling community at the university as well as by the precursory design of EngMeta. Within the repository, EngMeta serves as the semantic core and the repository is built around the metadata model, which is also deemed metadata-driven repository development. The requirements, such as handling large datasets, were stemming from aerodynamics and molecular dynamics [17].

DaRUS is based on Dataverse, and the driving factors for choosing this repository software were its design for research data management, its integration with the DOI persistent identifier infrastructure, its adaptability with metadata standards and its monolithic design. In the Dataverse repository software package, all the data is organized in Dataverses (organizational structure), datasets and files [18]. A Dataverse is the highest element in the hierarchical data organization structure in the repository and typically represents an institute or a research project. A dataset in the Dataverse terminology resembles to a directory or a collection of files. As of July 2020, DaRUS holds almost 600 files in 49 datasets, which are organized in 60 Dataverses, mainly from the fields of engineering, computer science and physics.

As DaRUS is an institutional repository, it is only loosely coupled to the research infrastructure since it is generic. This means that the service layer (l3) is basically the generic Dataverse web GUI. Additional services can be integrated by using one of the APIs that Dataverse offers, such as REST or SWORD. For example, an

---

[7]https://fordatis.fraunhofer.de/.

[8]https://darus.uni-stuttgart.de/.

automated toolchain (as an external tool) was implemented using the Dataverse API for the specific use case of thermodynamics: after a simulation run, an automated metadata extraction is triggered. Then, the extracted metadata altogether with the data is automatically ingested into the DaRUS repository [19].

2.2.3.2   NOMAD

In contrast to DaRUS, the Novel Materials Discovery (NOMAD) laboratory[9] (or Novel Materials Discovery Center of Excellence (NOMAD CoE)) is a prime example of a domain-specific data infrastructure which is highly integrated [20] in a virtual research environment. The repository part is complemented with the NOMAD Archive, the NOMAD Encylopedia, the NOMAD Visualization Tools and the NOMAD Analytics Toolkit. NOMAD is recommended by *Nature*[10] for depositing supplementary data when submitting a research article on materials modelling.

The NOMAD repository is the central component of the laboratory and holds input and output data from material simulations with a retention period of 10 years for free. The NOMAD archive holds the open-access data from the repository which was converted into a code-independent format. To accomplish this, developing a metadata definition and a metadata component was crucial for this. It serves, just as proposed in Sect. 2.1.1.1, as a common understanding[11] and, as the overall outline of this book, for making data semantically interoperable. The metadata definition uses 168 aligned and 2,360 code-specific metadata keys. For example, the different terms for quantities had to be mapped to one aligned term. According to [20], the development of this component of the data infrastructure was a challenge. The NOMAD encylopedia is the part of the NOMAD data infrastructure which provides millions of calculations via a web GUI with a materials-oriented view and therefore serves as knowledge base and a material classification system. The NOMAD visualization tools are a centralized service for data visualization within the data infrastructure allowing users interactive graphical analysis in materials modelling. Additionally, the NOMAD Analytics Toolkit is a big data analytics approach to support data evaluation, for example, scanning for specific thermoelectric materials or finding suitable materials for heterogeneous catalysis.

In the NOMAD laboratory, the archive and the repository components correspond to the storage layer (l1) and the object layer (l2), whereas the encyclopedia, the analytics toolkit and the visualization tools correspond to the service layer (l3), which is strongly coupled to the base layers.

As of February 2020, the NOMAD data infrastructure holds 49TB of raw data in the repository and 19TB of the archive in normalized, annotated form in 758 datasets.[12]

---

[9]https://www.nomad-coe.eu.

[10]https://www.nature.com/sdata/policies/repositories.

[11]https://www.nomad-coe.eu/the-project/nomad-archive/archive-meta-info.

[12]https://metainfo.nomad-coe.eu/nomadmetainfo_public/archive.html.

2.2.3.3 Materials Cloud

The Materials Cloud[13] is another domain-specific data infrastructure, which includes all the three aforementioned layers and implement them with specific technology supporting the data life cycle in materials modelling [11]. The Materials Cloud is, just like NOMAD, recommended by *Nature* for supplementary data for journal submissions in materials modelling. In the Materials Cloud, the ARCHIVE, DISCOVER, EXPLORE, WORK and LEARN components form according to data infrastructure.

The ARCHIVE component represents the open-access research data repository component with long-term storage, metadata protocols (including metadata harvesting for Google Dataset Search and B2FIND) and persistent identifiers (DOIs). The hardware backend of ARCHIVE is hosted at the Swiss National Computing Centre, is free of charge and data records are preserved for 10 years. For the software layer, Invenio will be used. ARCHIVE is moderated, which means all the ingested data is first checked against certain criteria, just as on preprint document servers. The DISCOVER component corresponds to the browsing capabilities for curated datasets of ARCHIVE and offers interactive visualization. The EXPLORE part of the system is the component that tracks and displays provenance information of the datasets to ensure FAIR and reproducible data. All this information is recorded by the AiiDA system, which can be imagined as a git style methodology for data. The information is shown in a provenance graph. The WORK component is the part of the Materials Cloud data infrastructure that allows working with the available data, which can be either stand-alone tools to perform inexpensive calculations or AiiDA lab. AiiDA lab is a tool for defining workflows and orchestrating them from the web interface, since it lets users connect and use remote computational resources or other repositories which include the OPTIMADE standard,[14] so, for example, NOMAD. The LEARN part of the system features educational material, such as tutorials or video lectures and a downloadable image of a virtual machine for training purposes in materials modelling. This part is important since it covers metadata processes as displayed in Sect. 2.1.1.4.

Just as NOMAD, the Materials Cloud is highly integrated data infrastructure, where the ARCHIVE component acts as the storage layer (l1) and the object layer (l2). The service layer (l3) is set up by DISCOVER, EXPLORE, WORK and LEARN components.

2.2.3.4 Chemotion, MoMaF and NFDI

The *Science Data Center for Molecular Materials Research (MoMaF)*[15] is one of the four Science Data Center (SDC) projects of the state of Baden-Württemberg in Germany started in late 2019. Its goal is to support the data life cycle and implement

---

[13]https://www.materialscloud.org/.

[14]https://www.optimade.org/.

[15]https://momaf.scc.kit.edu/.

the FAIR principles by a domain-specific repository for molecular materials research, digitalization of lab books and metadata standards.

MoMaF relies on preliminary work that was conducted in the Chemotion project,[16] whose aim was to build a data infrastructure for synthetic and analytic chemistry [21, 22]. The core of Chemotion is a repository that allows to collect, reuse and publish data. It is complemented with discipline-specific data processing tools and it incorporates DOI generation and supports publishing, such as support for peer-reviewing submissions and comparing submissions with the PubChem database. The repository architecture consists of a private workspace and a publication area. Electronic laboratory notebooks play a crucial role here and can be imported into the private workspace. Research data[17] can, after adding metadata and a reviewing process, later be staged from the private workspace to the publication area, where they are provided with a DOI and made Open Data. Also, within this approach, we can see how a repository on the object layer is complemented with additional tools in the service layer, such as data processing tools or electronic laboratory notebooks.

The work and the results from the MoMaF SDC will later be used in the National Research Data Infrastructure (NFDI) for Chemistry [23] as one of the NFDI projects in Germany. Another project within the NFDI, which also will have an impact for materials modelling, is NFDI for Catalysis.[18]

References

1. B. Schembera, D. Iglezakis, EngMeta: metadata for computational engineering. IJMSO **14**(1), 26–38 (2020)
2. A.J.G. Hey, A.E. Trefethen, The data deluge: an e-science perspective, in *Grid Computing: Making the Global Infrastructure a Reality*, ed. by F. Berman, G.C. Fox, A.J.G. Hey (Wiley, 2003), pp. 809–824
3. P.N. Edwards, M.S. Mayernik, A.L. Batcheller, G.C. Bowker, C.L. Borgman, Science friction: data, metadata, and collaboration. Soc. Stud. Sci. **41**(5), 667–690 (2011). https://doi.org/10.1177/0306312711413314
4. J. Greenberg, Metadata and the world wide web. Encycl. Libr. Inf. Sci. **3**, 1876–1888 (2003)
5. S. Staab, Wissensmanagement mit Ontologien und Metadaten. Inform-Spektr **25**(3), 194–209 (2002)
6. M.D. Wilkinson, M. Dumontier, J. Aalbersberg, G. Appleton, MA, et al., The FAIR guiding principles for scientific data management and stewardship. Sci. Data **3**, 160018 (2016)
7. EU, Turning FAIR into Reality (2018), https://ec.europa.eu/info/sites/info/files/turning_fair_into_reality_1.pdf, https://doi.org/10.2777/1524. Accessed 28 Apr 2019
8. B. Schembera, Forschungsdatenmanagement im Kontext dunkler Daten in den Simulationswissenschaften. Dissertation, Universität Stuttgart (2019). https://doi.org/10.18419/opus-11028
9. B. Schembera, J.M. Durán, Dark data as the new challenge for big data science and the introduction of the scientific data officer. Philos. Technol. **33**, 93–115 (2020)

---

[16]https://chemotion.net/.

[17]Chemotion has two structuring elements, which are samples (batches of a molecules) and reactions leading to the principle that information is kept along with and can be linked to the chemical process.

[18]http://gecats.org/NFDI4Cat.html.

10. B. Schembera, D. Iglezakis, The genesis of EngMeta: a metadata model for research data in computational engineering, in *Proceedings of MTSR 2018*, ed. by E. Garoufallou, F. Sartori, R. Siatri, M. Zervas, CCIS, vol. 8486 (Springer, Cham, Switzerland, 2018), pp 127–132

11. L. Talirz, S. Kumbhar, E. Passaro, A.V. Yakutovich, V. Granata, F. Gargiulo, M. Borelli, M. Uhrin, S.P. Huber, S. Zoupanos, C.S. Adorf, C.W. Andersen, O. Schütt, C.A. Pignedoli, D. Passerone, J. VandeVondele, T.C. Schulthess, B. Smit, G. Pizzi, N. Marzari, Materials cloud, a platform for open computational science. Sci. Data **7**, 299 (2020). arXiv:2003.12510 [cond-mat.mtrl-sci]

12. OAIS, Reference model for an Open Archival Information System. Technical Report 650.0-M-2 (Magenta Book) Issue 2, CCSDS (2012)

13. B. Schembera, T. Bönisch, Challenges of research data management for high performance computing, in *Proceedings of TPDL 2017*, ed. by J. Kamps, G. Tsakonas, Y. Manolopoulos, L. Iliadis, I. Karydis, LNCS, vol. 10450 (Springer, Heidelberg, Germany, 2017), pp. 140–151

14. J. Askhoj, M. Nagamori, S. Sugimoto, Archiving as a service: a model for the provision of shared archiving services using cloud computing, in *Proceedings of iConference 2011* (ACM, New York, USA, 2011), pp. 151–158

15. M. Hedges, A. Hasan, T. Blanke, Management and preservation of research data with iRODS, in *Proceedings of ACM 1st Workshop on CyberInfrastructure: Information Management in eScience*, ed. by P. Mitra (ACM, New York, USA, 2011), pp. 17–22

16. A. Wuchner, Das Projekt FORDATIS–Aufbau einer Forschungsdateninfrastruktur in der Fraunhofer-Gesellschaft, in *Forschungsdaten: sammeln, sichern, strukturieren, Forschungszentrum Jülich*, ed. by B. Mittermaier, pp. 57–78

17. B. Selent, H. Kraus, N. Hansen, B. Schembera, A. Seeland, D. Iglezakis, Management of research data incomputational fluid dynamics and thermodynamics. E-Science-Tage 2019–Data to Knowledge (2020). https://doi.org/10.11588/heibooks.598.c8422

18. B. Selent, B. Schembera, D. Iglezakis, A. Seeland, *Datenmanagement in Infrastrukturen* (Abschlussbericht des BMBF-Projektes DIPL-ING. Tech. rep., Universität Stuttgart, Stuttgart, Prozessen und Lebenszyklen für die Ingenieurwissenschaften, 2019). https://doi.org/10.2314/KXP:1693393980

19. B. Schembera, Like a rainbow in the dark: metadata annotation for HPC applications in the age of dark data. J. Supercomput. (2021). https://doi.org/10.1007/s11227-020-03602-6

20. C. Draxl, M. Scheffler, NOMAD: the FAIR concept for big data-driven materials science. MRS Bull. **43**(9), 676–682 (2018)

21. P. Tremouilhac, A. Nguyen, Y.C. Huang, S. Kotov, D.S. Lütjohann, F. Hübsch, N. Jung, S. Bräse, Chemotion ELN: an open source electronic lab notebook for chemists in academia. J. Chem. **9**(1), 1–13 (2017)

22. P. Tremouilhac, C.L. Lin, P.C. Huang, Y.C. Huang, A. Nguyen, N. Jung, F. Bach, R. Ulrich, B. Neumair, A. Streit, S. Bräse, The repository chemotion: infrastructure for sustainable research in chemistry. Ang. Chem. Int. Ed. (2020). https://doi.org/10.1002/anie.202007702

23. C. Steinbeck, O. Koepler, F. Bach, S. Herres-Pawlis, N. Jung, J.C. Liermann, S. Neumann, M. Razum, C. Baldauf, F. Biedermann, T.W. Bocklitz, F. Boehm, F. Broda, P. Czodrowski, T. Engel, M.G. Hicks, S.M. Kast, C. Kettner, W. Koch, G. Lanza, A. Link, R.A. Mata, W.E. Nagel, A. Porzel, N. Schlörer, T. Schulze, H.G. Weinig, W. Wenzel, L.A. Wessjohann, S. Wulle, NFDI4Chem: towards a national research data infrastructure for chemistry in Germany. Res. Ideas Outcomes **6**, e55852 (2020)

Chapter 3
Marketplace-Level Domain Ontologies

3.1 Ontologies and Formal Notation

This chapter and the subsequent two chapters present ontologies from the VIMMP project, their relation to other work (including other ontologies) and examples for their use in practice. While this is not a theoretical work, we begin with a brief introduction to the usual formal notation, on the one hand to support certain arguments, e.g. concerning ontology alignment, and on the other hand to make related literature, where such notation is employed, more accessible. For a dedicated presentation of Description Logic (DL), the logical formalism is employed for ontologies from the point of view of theoretical computer science, which exists in a great variety of versions, the reader is referred to Baader et al. [1] as well as Schneider and Šimkus [2]; the present work partly adheres to the notation used by the DL community, but deviates from it on occasion. Formal ontology also has philosophical aspects—which is natural given its origin in that discipline—that we do not address here; for this purpose, the reader is pointed to Berto and Plebani [3], whereas an easily accessible introduction to semantic technology from the point of view of ontology engineering is provided by Allemang and Hendler [4].

The formal representation of what is known in any given context is called a knowledge base; it is defined as a pair $\mathcal{K} = (\mathcal{T}, \mathcal{A})$, where $\mathcal{T}$ is the ontology and $\mathcal{A}$ is the scenario. In the DL community, $\mathcal{T}$ is called the TBox (terminological box) and $\mathcal{A}$ is called the ABox (assertional box)—hence the notation—while in model theory, $\mathcal{A}$ is referred to as a model. The ontology describes how we formalize a domain of knowledge, in general, irrespective of the circumstances, whereas the scenario contains statements that are contingent; depending on context this may be the meaning of the content of a database or a file. The ontology is given by a tuple $\mathcal{T} = (\mathbf{C}, \mathbf{R}, \Gamma)$, where $\mathbf{C}$ is a set of elementary concept (also "class" or "universal") names, $\mathbf{R}$ is a set of elementary relation (also "role") names, and Γ is a set of rules (also "general inclusions" or "axioms"). The scenario, in turn, is a tuple $\mathcal{A} = (\mathbf{I}, A_{\mathrm{c}}, A_{\mathrm{r}}, H)$, where $\mathbf{I}$ is a set of individual (also "object" or "particular")

© The Author(s) 2021
M. Horsch et al., *Data Technology in Materials Modelling*,
SpringerBriefs in Applied Sciences and Technology,
https://doi.org/10.1007/978-3-030-68597-3_3

names, A_c is a function representing conceptual assertions, such that an individual name $I \in \mathbf{I}$ is mapped to a set of concepts $A_c(I)$ and $A_r \subseteq \mathbf{I} \times \mathbf{R} \times \mathbf{I}$ is a set of relational assertions. The function H maps individuals $I \in \mathbf{I}$ to sets of elementary datatype property assertions $H(I) = \{\eta_1, \ldots, \eta_n\}$ where[1] each elementary datatype property assertion $\eta_i = (k_i, v_i)$ consists of a textual key $k_i \in \Sigma^\star$ and a textual or numerical value $v_i \in \mathbb{R} \cup \Sigma^\star$, with Σ representing the employed alphabet and $\Sigma^\star$ the free monoid over Σ.

On the semantic web, using the Web Ontology Language (OWL), elementary names correspond to Internationalized Resource Identifiers (IRIs), and anything referred to by an IRI, including concepts, relations and individuals, is accordingly also called a "resource". An IRI consists of a prefix and a suffix, where the alphabet Σ is the Unicode/ISO10646 universal coded character set. For the prefix, the same well-known syntax applies as for a URL, and indeed, resolvable locators can be used as IRIs; however, it is equally allowed to use non-resolvable identifiers, which cannot be directly looked up on the web. Most notations permit an abbreviation of the prefix so that osmo:workflow_graph can be written instead of the full IRI

https://purl.vimmp.eu/semantics/osmo/osmo.ttl#workflow_graph

for the concept name with the suffix workflow_graph from the ontology OSMO, cf. Sect. 3.3. Conceptual and relational assertions are of the type

$$I : C \qquad\qquad I \text{ a } C,$$
$$(I, J) : R \qquad\qquad I \ R \ J, \qquad\qquad (3.1)$$

respectively, where $I, J \in \mathbf{I}$ are individuals, C is a concept and R is a relation (in OWL, an owl:ObjectProperty)[2]; the assertions above state that $C \in A_c(I)$ and $(I, R, J) \in A_r$. DL notation [1] is given on the left side and Terse Triple Language (TTL) notation [4] on the right side; "triple" here refers to the sequence of a subject (e.g. I), a predicate (e.g. C) and an object (e.g. J). The predicate "a" in TTL notation is an abbreviation for rdf:type, referring to the Resource Description Framework (RDF), so that I a C indicates that C is the type of I, i.e. that I is an individual that instantiates the concept C.

Examples for triples corresponding to conceptual and relational assertions, using OSMO (Sect. 3.3) and VOV (Chap. 4), would be the following rendering of "ex:D is a dipole moment vector and its value is zero:"

ex:D	a	vov:electric_dipole_moment;
	vov:shares_value_with	osmo:ZERO_VECTOR_3D. (3.2)

---

[1] Dedicated description logics that were developed to formalize elementary datatype properties (owl:DatatypeProperty) include DAML+OIL [5–7] and $\mathcal{SHOQ}(\mathbf{D})$ [8].

[2] Elementary datatype properties $(k, v) \in H(I)$ are asserted as $I k v$ in TTL, where the key k is an owl:DatatypeProperty.

In TTL notation, ending the first triple with a semicolon implies that the subject (here, ex:D) is reused for the second triple, the full representation of which would therefore be ex:D vov:shares_value_with osmo:ZERO_VECTOR_3D. The DL notation for Expression (3.2) is

$$\text{ex:D} \; : \; \text{vov:electric\_dipole\_moment,}$$
$$(\text{ex:D, osmo:ZERO\_VECTOR\_3D}) \; : \; \text{vov:shares\_value\_with.} \qquad (3.3)$$

Operators can be applied to elementary names, yielding composites such as[3]

$C \sqcap C'$	[owl:intersectionOf ($C\,C'$)],
$C \sqcup C'$	[owl:unionOf ($C\,C'$)],
$\neg C$	[owl:complementOf C],
R^-	[owl:inverseOf R],
$\exists R.C$	[a owl:Restriction; owl:onProperty R; owl:someValuesFrom C],
$\forall R.C$	[a owl:Restriction; owl:onProperty R; owl:allValuesFrom C],
$C^\bullet$	[rdfs:domain C],
$^\bullet C$	[rdfs:range C], $\qquad\qquad\qquad\qquad\qquad\qquad$ (3.4)

where C and C' are concepts; R is a relation; and $\sqcap$, $\sqcup$ and $\neg$ denote the intersection, union and complement of concepts, respectively, and R^- denotes the inverse relation to R. For an individual $I \in \mathbf{I}$, the assertion $I : \exists R.C$ entails that there is a $J \in \mathbf{I}$ with $J{:}C'$ and $(I, J) : R$, whether explicitly asserted or not, while $I : \forall R.C$ entails $J{:}C'$ for all $J \in \mathbf{I}$ with $(I, J) : R$, whether explicitly asserted or not. The relation $C^\bullet$ holds whenever its subject (i.e. its first argument) is an individual that instantiates C, i.e. $C^\bullet$ relates all C individuals to all individuals; $^\bullet C$ holds whenever its object (i.e. its second argument) instantiates C, i.e. $^\bullet C$ relates all individuals to all C individuals.

Rules can include subsumption and equivalence for concepts and relations[4]

---

[3]In the last two rows, for $C^\bullet$ and $^\bullet C$, the TTL version is an approximation, since in OWL, the use of domain and range composites is restricted to what would best be represented formally as a subsumption rule, cf. Expression (3.5). The subsumption $R \sqsubseteq C^\bullet$ then becomes R rdfs:domain C, and $R \sqsubseteq {}^\bullet C$ becomes R rdfs:range C. However, it is impossible to use this construction for any other purpose, e.g. to state $R \equiv C^\bullet$.

[4]The use of the article "a" is a possible source of misunderstandings between communities due to the way in which it is treated by two different notations, namely, TTL, which is used for the VIMMP ontologies and throughout this book, as opposed to the Open Biomedical Ontologies (OBO) format [9]. TTL uses a for instantiation (rdfs:type) such as in macro:VIMMP_MARKETPLACE a macro:digital_marketplace, signifying "the VIMMP Marketplace is a digital marketplace." OBO format denotes conceptual subsumption (rdfs:subClassOf) by the keyword is_a, such as in id: macro:digital_marketplace [...] is_a: macro:bidirectional_channel, signifying "every digital marketplace is a bidirectional channel," since macro:digital_marketplace $\sqsubseteq$ macro:bidirectional_channel. Motivated by that standard, the OWLViz tool [10] automatically labels all arrows with "is-a" when visualizing a taxonomy. Where this occurs here (Figs. 3.2, 3.3, etc.), it should be understood as "is subclass of".

$$C \sqsubseteq C' \qquad\qquad C \text{ rdfs:subClassOf } C',$$
$$R \sqsubseteq R' \qquad\qquad R \text{ rdfs:subPropertyOf } R',$$
$$C \equiv C' \qquad\qquad C \text{ owl:equivalentClass } C',$$
$$R \equiv R' \qquad\qquad R \text{ owl:equivalentProperty } R', \qquad (3.5)$$

such as

osmo:simulation_workflow

$$\sqsubseteq \text{ osmo:workflow\_graph } \sqcap \exists S.\text{evmpo:simulation}, \qquad (3.6)$$

where S is an abbreviation for the relation "is sign for" (viprs:is_sign_for, cf. Sect. 5.2). Accordingly, Expression (3.6) states that "every simulation workflow is a workflow graph that is a sign for a simulation", relating the concept of a simulation workflow to those of a workflow graph and a simulation.

Similarly,

osmo:is_governing_equation_in

$$\sqsubseteq \text{ P } \sqcap \text{ osmo:governing\_equation}^\bullet \sqcap {}^\bullet\text{osmo:materials\_model}, \qquad (3.7)$$

where P denotes "is proper part of" (viprs:is_proper_part_of) states that "if I is a governing equation in J, then I is a governing equation, J is a materials model and I is a proper part of J." In TTL notation, this is expressed as

osmo:is_governing_equation_in	rdfs:domain	osmo:governing_equation;
	rdfs:range	osmo:materials_model;
	rdfs:subPropertyOf	viprs:is_proper_part_of.

Other types of rules concern the disjointness of concepts and algebraic properties of relations such as symmetry, transitivity and reflexivity.

Different types and fragments of DL restrict composites and rules that can be included in a knowledge base in various ways to avoid computational undecidability, and beyond this, to limit the complexity of reasoning tasks [1]. This is also the case for OWL DL, the description logic associated with OWL as well as the DL language profile of OWL2, which is the main standard for ontology engineering [11]. Adherence to the expressivity restrictions of this logic is prescribed by reasoners such as FaCT++ and other widespread tools such as *protégé*. Relational composites ($R \sqcap R'$, etc.) cannot be included as such[5] in OWL DL; however, indirect constructions can often be devised. Chain relations of the type $R_1 \circ R_2$, with the usual meaning

$$(I, J) : (R_1 \circ R_2) \iff \exists I' \in \mathbf{I} : (I, I') : R_1 \wedge (I', J) : R_2 \qquad (3.8)$$

---

[5] We will use such notational constructions here nonetheless, where appropriate, with the intuitive meaning, e.g. $(I, J) : (R \sqcap R') \iff (I, J) : R \wedge (I, J) : R'$.

may be used for rules, but only under a relatively complex set of constraints; for details, cf. the reference manual on OWL2 language profiles [12] as well as the textbooks mentioned above [1, 4].

The present discussion of ontologies in materials modelling will generally limit itself to the minimum required level of theoretical detail. Section 3.2 introduces the European Virtual Marketplace Framework, constituted by multiple Horizon 2020 projects, and its (comparably small and abstract) ontology EVMPO; on this basis, Sects. 3.3 to 3.5 introduce the modelling service-oriented marketplace-level domain ontologies MACRO, MMTO, OSMO, OTRAS, VICO and VIVO, and Chap. 4 introduces the marketplace-level domain ontologies VISO and VOV that permit the description of simulation codes, materials models and associated quantities such as model parameters and thermodynamic properties or boundary conditions. Section 5 discusses top-level ontologies, which are at the highest level of abstraction (and therefore rather formal and philosophical), the alignment between top-level and domain ontologies, and practical applications from CME/ICME-based process data technology.

3.2 European Virtual Marketplace Framework

To enable semantic interoperability and FAIR data management,[6] the VIMMP project has developed a system of marketplace-level domain ontologies, cf. Fig. 3.1, supporting the ingest and retrieval of data and metadata at the VIMMP marketplace front end [14]; these ontologies are expressed in OWL2 using TTL notation [4]. Internally, VIMMP uses the marketplace-level domain ontologies as a part of its approach to data management, underlying the interactions with users at its front end [14]. VIMMP contributes to the activities of the EMMC to coordinate these developments with the community and the ecosystem of platforms developed from related projects funded from the Horizon 2020 research and innovation programme.

The *European Virtual Marketplace Framework* (EVMF) establishes an ecosystem of interoperable environments that builds on previous EMMC standardization efforts, including RoMM [17], the EMMC Translation Case Template [18], the EMMC Translators' Guide [19] and the MODA metadata standard for simulation workflows [20]. Within this interoperability framework, any provider will have the possibility to choose the depth at which any provided services and tools implement jointly agreed semantics: the deeper the adherence, the deeper the capability to interoperate with other platforms and services. While the EVMF was established by a collaboration between the VIMMP and MarketPlace consortia in coordination with the EMMC, it is open to participation by all developers, providers, translators and end users of services in materials modelling. The EVMF is entirely based on transpar-

---

[6]Disclaimer: Contents from Sects. 3.2–3.5, 4.3, 4.4, 5.2 and 5.4 are also included in the openly accessible documentation of the VIMMP ontologies [13], which is distributed with the VIMMP ontology release.

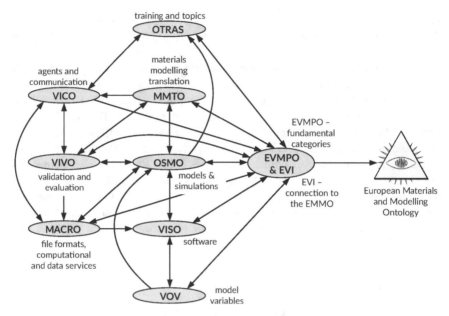

Fig. 3.1 *Ellipses: Ontologies developed by VIMMP [14]. Triangle: EMMO, the employed top-level ontology [15, 16], used here in combination with VIPRS (cf. Sect. 5.2). An arrow signifies that an ontology refers to concepts or relations from another ontology*

ent and openly accessible specifications, relying on the EMMO at the top level [15, 16]; the present ontologies are accordingly released as free software under the GNU Lesser General Public License (LGPL). By creating an open framework on the basis of community-governed interoperability standards, a variety of projects, several of which (including VIMMP, MarketPlace and OntoCommons) are funded from Horizon 2020—or prospectively may be funded from the Horizon Europe programme in the future, contribute to a system of platforms and infrastructures that will support the uptake of materials modelling solutions by industrial research and development practice.

The *European Virtual Marketplace Ontology* (EVMPO) was developed by the VIMMP, MarketPlace and EMMC-CSA projects as a common point of departure for the standardization of service-oriented semantics, with a focus on digital marketplace platforms in materials modelling.[7] By defining 11 *fundamental paradigmatic categories*, which correspond to irreducible terms that are constitutive to the paradigm underlying materials modelling marketplaces, the EVMPO provides a basic structure for the development of marketplace-level domain ontologies. The fundamental paradigmatic categories are defined as follows:

---

[7]EVMPO: `https://emmc.eu/semantics/evmpo/evmpo.ttl` (non-resolvable IRI), mirrored at `http://www.molmod.info/semantics/evmpo.ttl` (resolvable URL).

1. evmpo:assessment: a proposition on the accuracy or performance of an entity or an a expression of trust in an entity. Corresponding domain ontology: VIVO, cf. Sect. 3.4.
2. evmpo:calendar_event: a meeting or activity which is scheduled or can be scheduled; this is defined to be equivalent with Vevent from the W3C iCalendar ontology (iCal) with time zones as datatypes, cf. Connolly and Miller [21]. Corresponding domain ontology: OTRAS, cf. Sect. 3.5.
3. evmpo:communication: any message (or an attachment or part of a message) that is communicated. Corresponding domain ontology: VICO, cf. Sect. 3.5.
4. evmpo:information_content_entity: a journal article, a dataset or a graph. This concept is defined to be equivalent with IAO_0000030, labelled "information content entity" through rdfs:label, from the Information Artifact Ontology (IAO), cf. Ceusters [22]. Corresponding domain ontologies: OTRAS and VISO, cf. Sect. 3.5 and Chap. 4.
5. evmpo:infrastructure: infrastructure of an EVMF-interoperable platform (e.g. related to data, hardware and software). Corresponding domain ontologies: MACRO and VISO, cf. Sect. 3.3 and Chap. 4.
6. evmpo:interpreter: this concept is defined to be the same as emmo-semiotics:Interpreter from the nominalist revision of Peirce's semiotics, based on the semiotic triad sign—object—interpretant, as included in the EMMO [15, 16]; therein, for any given triad, the interpreter is the entity that carries out the semiosis, taking the sign (a representamen) as an input and producing the interpretant (another representamen) as an output [16, 23, 24]. Therefore, any potential agent or communicating entity at EVMF-interoperable infrastructures is an interpreter. Corresponding domain ontology: VICO, cf. Sect. 3.5.
7. evmpo:material: an amount of a physical substance (or mixture of substances) that is part of a more comprehensive real-world object; this concept is defined to be the same as emmo-physicalistic:Material from the EMMO [15, 16]. Corresponding domain ontologies: OSMO and VIVO, cf. Sects. 3.3 and 3.4.
8. evmpo:model: a sign that represents a physical object or process by direct similitude and/or within a mathematical framework; this concept is defined to be the same as emmo-models:Model from the EMMO [15, 16]. Corresponding domain ontologies: OSMO, VISO and VOV, cf. Sect. 3.3 and Chap. 4.
9. evmpo:process: the temporal evolution of one or multiple entities. Corresponding domain ontologies: MMTO, OSMO and VISO, cf. Sects. 3.3 and 3.4 as well as Chap. 4.
10. evmpo:product: a good or service, to be acquired either on a EVMF-interoperable digital marketplace or off-site; services that can be traded on a digital marketplace are specifically conceptualized by evmpo:tradeable_object. Corresponding domain ontologies: MACRO, MMTO and OTRAS, cf. Sects. 3.3, 3.4 and 3.5.
11. evmpo:property: a representamen that is determined as an interpretant from an observation process, involving a specific observer that perceives or measures it; this concept is defined to be the same as emmo-properties:Property

from the EMMO [15, 16]. Corresponding domain ontologies: VIVO and VOV, cf. Sect. 3.4 and Chap. 4.

These categories need not be disjoint, e.g. evmpo:material and evmpo:product overlap, since a material can be manufactured with the intent of selling it as a commodity, by which it becomes a good, and hence a product.

The common superclass of the paradigmatic categories is evmpo:paradigmatic_entity. Below the fundamental level, the EVMPO also includes non-fundamental entites as subclasses, e.g. evmpo:simulation as a subclass of evmpo:process and evmpo:service as a subclass of evmpo:product. Terms which are not closely related to the materials modelling marketplace paradigm itself, but may occur within a related knowledge base, are defined to be *non-paradigmatic*. For this purpose, the EVMPO includes evmpo:annotation as a twelfth, non-paradigmatic fundamental category; the EVMPO top relation, parent to both evmpo:paradigmatic_entity and evmpo:annotation, is evmpo:marketplace_related_entity. The relation evmpo:has_annotation can connect any marketplace-related entity to an annotation. Below this, 12 subproperties are defined, corresponding to the fundamental categories, i.e. evmpo:has_assessment_annotation pointing to annotations of an assessment, etc., and evmpo:has_meta_annotation for annotations of an annotation.

Consistency with the EVMPO, and by implication consistency with the EMMO, is a requirement for all components and infrastructures that aim at interoperating within the EVMF. This design ensures that while EVMF-interoperable infrastructures need to agree on the definition of the most important entities, any platform retains the option to extend its own semantic base as required. To remain interoperable within the EVMF, any additional concepts need to be subsumed under fundamental categories from the EVMPO; cf. Sect. 5.4 for a summary on how the EVMPO and the marketplace-level domain ontologies are aligned with the EMMO.

3.3   Modelling, Simulation and Computational Resources

The *Marketplace-Accessible Computational Resource Ontology* (MACRO) deals with data and hardware-related resources and infrastructures [14]. In particular, MACRO contains classes and individuals representing file formats expected to occur on the VIMMP marketplace platform,[8] many of which are obtained by connecting to the EDAM ontology [25]. High-level concepts from MACRO and their relation to EVMPO concepts (evmpo:agent, evmpo:annotation, etc.) are shown in Fig. 3.2. Complementing MACRO, the PaaSPort ontology [26] can be used to describe Platforms as a Service (PaaS).

The *Ontology for Simulation, Modelling and Optimization* (OSMO) was developed as the ontology version of MODA [20], making workflow representations

---

[8]MACRO: https://purl.vimmp.eu/semantics/macro/macro.ttl (non-resolvable IRI), mirrored at http://www.molmod.info/semantics/macro.ttl (resolvable URL).

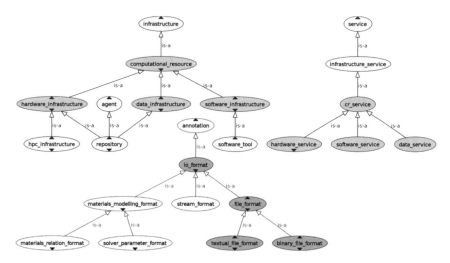

Fig. 3.2 Annotation, infrastructure and service branch fragments of the MACRO class hierarchy, version 1.1.4. The OWLViz protégé plugin was used to visualize the ontology [10]; arrows labelled "is-a" denote subsumption ($\sqsubseteq$)

machine processable,[9] semantically interoperable with community platforms and amenable to automated reasoning [27]. Where a physics-based modelling approach is followed, Physical Equations (PEs) are employed jointly with Materials Relations (MRs) that parameterize and complement the PEs, e.g. for a particular substance. The combination of PEs and MRs is referred to as the system of governing equations; on the basis of RoMM [17], common PE types are subdivided into four groups according to their granularity level: electronic, atomistic, mesoscopic and continuum [17, 20, 27]. In MODA graphs, there are four types of vertices (corresponding to Sects. 3.1–3.4 of the MODA form), which are in OSMO referred to as sections (osmo:section):

1. Use case (osmo:use_case)—MODA Sect. 3.1. The physical system to be simulated, including information on the given and desired physical properties. In OSMO, the application case (osmo:application_case) is introduced as a more general concept, permitting the description of applications of the simulation outcome that go beyond the immediate simulation scenario [28].

2. Model—MODA Sect. 3.2. The system of GEs, with one or multiple PEs and MRs; here, this is referred to as a materials model (osmo:materials_model). Following the EMMO approach, implemented by the EVMPO, a model (evmpo: model) is conceptualized, substantially more broadly, as an icon (representamen) providing a simplified representation of a physical object that is suitable for predicting its behaviour [15, 16].

---

[9]OSMO: `https://purl.vimmp.eu/semantics/osmo/osmo.ttl` (non-resolvable IRI), mirrored at `http://www.molmod.info/semantics/osmo.ttl` (resolvable URL).

3. Solver (osmo:solver)—MODA Sect. 3.3. The numerical solution of the model—defined with a strict limitation to considering exactly the variables that occur in the GEs explicitly (and nothing else).
4. Processor (osmo:processor)—MODA Sect. 3.4. Any computational operation beyond the above; in particular, this includes and processing activity done by a simulation code that goes beyond the immediate solution of underlying governing equations, e.g. to produce aggregated output. MODA is—strictly speaking—limited to postprocessors. Depending on the role in the simulation workflow, OSMO distinguishes between preprocessor, postprocessor, coupled (i.e. synchronous) and data processor elements.

For each section, the MODA standard contains a list of text fields, which are here referred to as section aspects, through which detailed information can be provided. In OSMO, the detailed description of section individuals by section aspects and their textual, numerical or object content is closely aligned with the corresponding textual and numerical entries from MODA; by using the relation osmo:has_aspect_object_content, it becomes possible to point to content provided anywhere on the semantic web, including individuals and classes from the VIMMP marketplace-level domain ontologies. Providing a common semantic basis for workflows, cf. Fig. 3.3, OSMO can be employed to consistently integrate data provenance descriptions for materials modelling data from diverse sources [27].

Selected concepts from MACRO and OSMO:

- macro:channel: a data infrastructure which, in its evolution as a process, contains communication events (semioses).
- osmo:condition: a statement concerning values of properties and/or parameters and/or their relation to each other. Subclasses include mmto:kpi_model.
- osmo:einecs_listed_material: an EC listed material from the *European Inventory of Existing Commercial Chemical Substances* (EINECS), which can be identified by an EC number; analogous: osmo:cas_listed_material, identified by a CAS number.
- macro:io_format: a syntactical convention to which a technical I/O implementation can adhere.
- osmo:logical_variable: a term that can be exchanged by interaction with logical resources. Subclasses include osmo:unique_elementary (for scalar variables) and osmo:optimization_objective.
- osmo:materials_relation: a Materials Relation (MR) as defined by RoMM [17], cf. MODA entry 2.4 [20].
- macro:model_database: a repository that can act as a model provider.
- osmo:section_aspect: a descriptor of a section (osmo:section), following the approach from MODA [20].
- osmo:workflow_graph: an LDT workflow graph-based description of a simulation, or a part of such a graph-based description.

Selected relations from MACRO and OSMO:

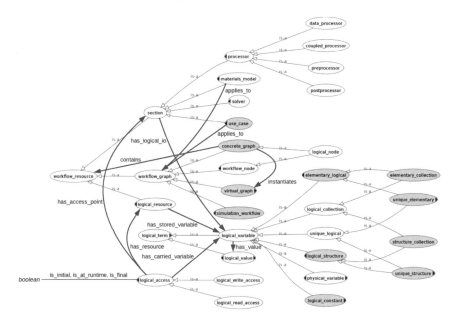

Fig. 3.3 Workflow resource branch fragment of the OSMO class hierarchy, version 1.6.8. The OWLViz protégé plugin was used to visualize the ontology [10]; arrows labelled "is-a" denote subsumption (⊑), and the other arrows denote relations; solid line: datatype properties (flags) associated with the osmo:logical_access concept

- osmo:has_aspect points to an aspect associated with a section. Domain: osmo:section; range: osmo:section_aspect.
- osmo:has_aspect_object_content points to an object entry associated with an aspect. Domain: osmo:section_aspect; range: evmpo:marketplace_related_entity.
- macro:has_channel_member points to an agent that participates in communicating through a channel. Domain: macro:channel; range: evmpo:agent.
- macro:has_granularity points to the granularity level to which the entities represented in an I/O format belong. Domain: macro:materials_modelling_format; range: osmo:granularity_level.
- osmo:has_value points to a value assigned to a logical variable. Domain: osmo:logical_variable; range: osmo:logical_value.
- osmo:has_variable_unit points to the unit to be associated with any assigned decimal values. Domain: osmo:elementary_logical; range: vivo:unit.
- macro:is_io_format_of points to a software tool that can process files in a given I/O format. Domain: macro:io_format; range: viso:software_tool.
- osmo:is_linked_to. (:F is linked to :G) ⟺ :F and :G cannot be executed concurrently—one side depends on the completion of the other side. Domain: osmo:workflow_graph; range: osmo:workflow_graph.

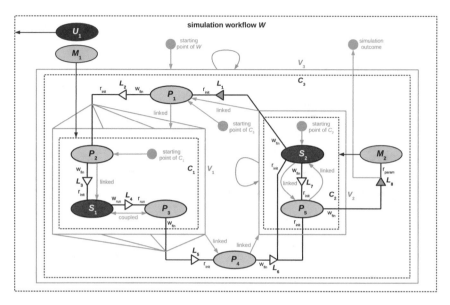

Fig. 3.4 Example simulation workflow in LDT notation; scenario: Molecular-simulation-based automated parameterization of a phenomenological equation of state [27, 29]

- macro:provides_access_to points to a service that can be accessed through the given infrastructure. Domain: macro:infrastructure; range: macro:infrastructure_ service.

Logical Data Transfer (LDT) notation [27] builds on the workflow graph representations from MODA [20]; it clarifies how exactly the use case, model, solver and processor entities relate to each other, e.g. where iterations and user interactions take place and how variables are exchanged. In LDT notation, cf. Fig. 3.4, ellipses represent sections; green circles and green arrows represent coupling and linking of elements, dependencies concerning the order of execution and aspects related to concurrency and synchronization. Blue arrows point from use cases and models to the part of the workflow to which these elements apply. Triangles are logical resources, describing how information is transferred between the sections, pointing from the source to the destination; if a triangle is filled, this denotes that a user interaction can occur.

The visualization elements from LDT notation have a direct correspondence with concepts and relations from OSMO [27], e.g. coupling and linking symbolized by green arrows correspond to the relations osmo:is_coupled_with and osmo:is_linked_to, and flow of information, represented by lines between triangles ellipses, corresponds to osmo:logical_access entities that relate to a logical resource by osmo:has_resource and to a section by osmo:has_access_point, cf. Fig. 3.3. The LDT representation, therefore, corresponds to an enriched version of a MODA graph; by removing logical resources, details on iterations (represented in OSMO by relations between "virtual graphs" and "concrete graphs"), etc., a conventional MODA description can be obtained. Similarly, the usual human-readable MODA forms can be obtained by reducing all OSMO aspects to an elementary numerical or textual description.

3.4 Engineering Applications and Validation

The *Materials Modelling Translation Ontology* (MMTO) deals with the process of "translating" a problem from engineering practice to modelling and simulation— and from the simulation outcome back to an actionable decision [28]. The role of the materials modelling translator is specified in detail by the EMMC Translators' Guide (ETG), cf. Hristova et al. [19]; accordingly, a translator needs to be able to bridge the "language gap" between industrial end users and academic model providers and software owners. The work of a translator aims at delivering not just modelling results, but a solution for an industrial engineering problem, understood more holistically. In business administration and management, such problems are usually addressed in terms of Key Performance Indicators (KPIs), where a KPI is understood to be a descriptor (indicator) underlying process and product opimization, ultimately characterizing some feature or property that can serve as a selling argument. The underlying orientation towards marketing reflects a point of view corresponding to organizational roles that are comparably distant from research and development. In scenarios that arise in such a context, it necessarily appears to be most crucial to address concerns that are immediately relevant to Business-to-Administration (B2A), Business-to-Business (B2B) and Business-to-Customer (B2C) relations [30].

In the MMTO,[10] which predominantly targets communities of users in engineering practice (rather than industrial business administration), the concept mmto:key_performance_indicator is reserved for scalar quantities that are relevant for characterizing, modelling or optimizing processes and products by CME/ICME methods. On this basis, two major distinctions are to be made from the point of view of a materials modelling translator [28]:

1. Some KPIs are closely related to human sentience (aesthetics, haptics, taste, etc.). Studies aiming at gaining information on these quantities typically rely on market research and other empirical methods that involve human subjects; such indicators are referred to as subjective KPIs (mmto:subjective_kpi). Obversely, an objective KPI (mmto:objective_kpi) can be determined by a standardized process, e.g. a measurement, experiment or simulation, the result of which (assuming that it is conducted correctly) does not depend on the person that carries it out.
2. An objective KPI is technological (mmto:technological_kpi) if it is observed or measured within a technical or experimental process, referring directly to proper-

---

[10]MMTO: https://purl.vimmp.eu/semantics/mmto/mmto.ttl (non-resolvable IRI), mirrored at http://www.molmod.info/semantics/mmto.ttl (resolvable URL).

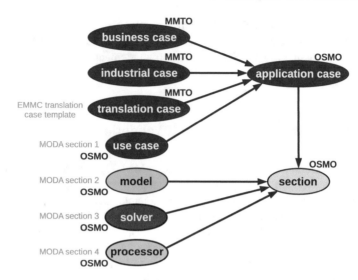

Fig. 3.5 Section branch fragment of the MMTO (version 1.3.4) and OSMO (version 1.6.8) class hierarchies; arrows denote subsumption (⊑)

ties of the real product or manufacturing process; properties of a model, which are determined by simulation, are computational KPIs (mmto:computational_kpi).

The distinction between subjective and objective KPIs is similar to that between Critical-to-Customer (CTC) and Critical-to-Quality (CTQ) measures [31–33]. The formulation given above, however, is more closely related to concepts from the EMMO. Due to the underlying approach to semiotics [15, 16, 23, 24], it is straightforward in the EMMO to categorize signs by the way in which their interpretation depends on the subjective impression of an interpreter or observer: in particular, the same distinction between "subjective properties" and "objective properties" is made in the EMMO; accordingly, the present approach supports a straightforward alignment of the MMTO with the EMMO and the approach to interoperability guided by the EMMC and implemented within VIMMP.

An instance of the materials modelling translation process, some agreed features of which are codified by the ETG and the EMMC Translation Case Template (ETCT) [18], is referred to as a Translation Case (TC). According to these specifications, a materials modelling translation project begins with exploring and understanding the Business Case (BC) and the Industrial Case (IC), or multiple relevant BCs and/or ICs, which characterize socioeconomic objectives and boundary conditions [28]. Universals for BCs (mmto:business_case), ICs (mmto:industrial_case) and TCs (mmto:translation_case) are defined to be subclasses of (osmo:application_ case), by which they can be dealt with in a similar way as the sections from OSMO; the class hierarchy of the section branch of the MMTO and OSMO is visualized in Fig. 3.5. In this way, the MMTO generalizes this approach from MODA to also cover the translation-related concepts from the ETCT and the ETG [18, 19, 28]. The TC aspects directly correspond to the ETCT text fields [18], except that as an ontology, the MMTO permits the incorporation of semantically characterized content from the semantic web. A business case can represent any purely economic consideration or an optimization problem at the management level, whereas an IC refers to an industrial engineering problem or an optimization problem at the technical or research and development level. Within the translation process, a suitable approach based on modelling and simulation is identified and carried out; subsequently, the outcome is translated back to support an actionable decision at the BC and IC levels. Thus, the MMTO is also a tool for representing exchange of information during translation processes (e.g. employing KPIs as logical variables), which may be represented by workflow graphs following the MODA and/or LDT workflow notations [20, 27].

The *VIMMP Validation Ontology* (VIVO) categorizes assessments (i.e. evaluations) of computational resource requirements and benchmarking as well as customer feedback on various kinds of entities,[11] which can be provided subsequent to transactions at the VIMMP marketplace [14]. Thereby, users support each other mutually, evaluating contents and providers, while the marketplace platform itself remains neutral and equally open and accessible to everybody. A matrix with subclasses of evmpo:assessment, indicating how marketplace users can evaluate what sort of objects, is shown in Fig. 3.11. Rows correspond to classes of entities that are subjected to an assessment such that a vivo:data_infrastructure_assessment is an evmpo:assessment that vivo:evaluates an evmpo:data_infrastructure, and a vivo:meta_assessment is an evmpo:assessment that vivo:evaluates an evmpo:assessment. Columns correspond to different ways in which entities can be evaluated, e.g. by reporting an observation on the relative quantitative accuracy (vivo:relative_accuracy_assessment) or by issuing a recommendation (evmpo:endorsement_assessment). Not all theoretically conceivable combinations are allowed—e.g. memory requirements can be stated for software, but not for projects. Using VIVO, in particular, error analyses and estimates can be attributed to models, simulation workflows and to data items obtained from repositories or other platforms (Fig. 3.6).

Selected concepts from the MMTO and from VIVO:

- vivo:assertion: a claim or proposition (e.g. as part of an assessment). Subclasses include vivo:accuracy_assertion, evmpo:material_property_information and vivo: requirement_assertion.
- mmto:business_decision_support_system: a decision support system that is applied to a business case. Analogous: mmto:industrial_decision_support_system, mmto:translation_decision_support_system.
- vivo:certificate: a validation statement by which an assessment is stated.
- vivo:computational_time_requirement: a requirement assessment concerning the computational (CPU time) requirements of a simulation workflow.

---

[11]VIVO: https://purl.vimmp.eu/semantics/vivo/vivo.ttl (non-resolvable IRI), mirrored at http://www.molmod.info/semantics/vivo.ttl (resolvable URL).

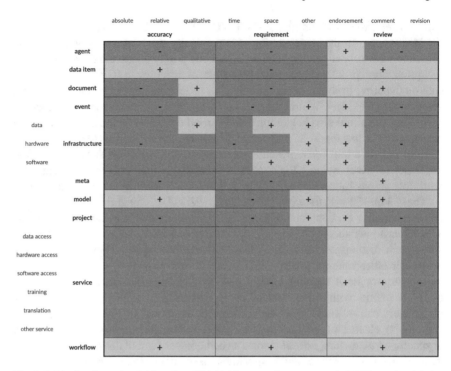

Fig. 3.6 Matrix of permitted (+) and prohibited (−) types of assessments in VIVO version 1.1.4

- vivo:material_property_information: an assertion referring to a material property.
- mmto:objective_kpi: a KPI that can be determined by a standardized process, the outcome of which is expected not to depend on the interpreter.
- vivo:relative_deviation: an accuracy assertion in which the relative magnitude of an error or uncertainty is given, normalized by the absolute magnitude of the value to which the assertion refers. Analogous: vivo:absolute_deviation.
- mmto:translation_case: an application case that can be described as specified by the ETCT [18].
- vivo:unit: a unit that can be expressed as a scalar multiple of an algebraic combination of SI units. This concept is the same as emmo-metrology:ReferenceUnit and qudt:Unit [15, 16, 34].

Selected relations from the MMTO and from VIVO:

- mmto:considers_business_case points to a business case considered within translation stage no. 1, "good understanding of the business case," as specified in the ETG [19]. Domain: mmto:translation_step_bc; range: mmto:business_case. Analogous: mmto:considers_industrial_case, corresponding to the translation stage no. 2, "good understanding of the industrial case" [19].

- mmto:describes_product points to a product that is described with the given KPI. Domain: mmto:key_performance_indicator; range: evmpo:product. Analogous: mmto:describes_process.
- vivo:evaluates points to the object evaluated by an assessment. Domain: evmpo:assessment; range: evmpo:marketplace_related_entity.
- vivo:has_assertion points to an assertion made within an assessment. Domain: evmpo:assessment; range: vivo:assertion.
- vivo:has_error_statement points to an accuracy assertion contained within a material property information. Domain: vivo:material_property_information; range: vivo:accuracy_assertion.
- mmto:has_tca_pe_type points to a TC aspect (as included in the ETCT) containing information on PE types employed during modelling. Domain: mmto:translation_case; range: mmto:tca_pe_type.
- vivo:has_unit points to the unit in which any numerical contents of an assertion are given. Domain: vivo:assertion; range: vivo:unit.
- vivo:is_quantity_kind points to the physical property characterization following QUDT [34]. Domain: vivo:assertion; range: qudt:QuantityKind.
- vivo:states_assessment points to an assessment contained within a certificate. Domain: vivo:certificate; range: evmpo:assessment.

3.5 Training and Communication

The *Ontology for Training Services* (OTRAS) can be employed to annotate any training resources in the field of materials modelling [14], i.e. [12] both training documents (such as manuals or videos) and training events (lectures, seminars, summer schools, workshops, etc.). In OTRAS, such resources are referred to as *carriers*. For information on training courses, syllabi, etc., the *Course Curriculum and Syllabus Ontology* (CCSO) is employed [35]. Furthermore, the IAO is applied to documents [22], in accordance with the EVMPO. The high-level structure of OTRAS is shown in Fig. 3.7. While the CCSO covers much of the required domain at an abstract level, a dedicated standardization effort is required to characterize the semantic space with respect to training contents specifically in the field of materials modelling. For this purpose, OTRAS includes a formalism by which learning outcomes and expert competencies can be described and a taxonomy of topics in materials modelling. Concerning didactics, the normal form of a learning outcome description to be used with OTRAS is given as follows:

---

[12]OTRAS: https://purl.vimmp.eu/semantics/otras/otras.ttl (non-resolvable IRI), mirrored at http://www.molmod.info/semantics/otras.ttl (resolvable URL).

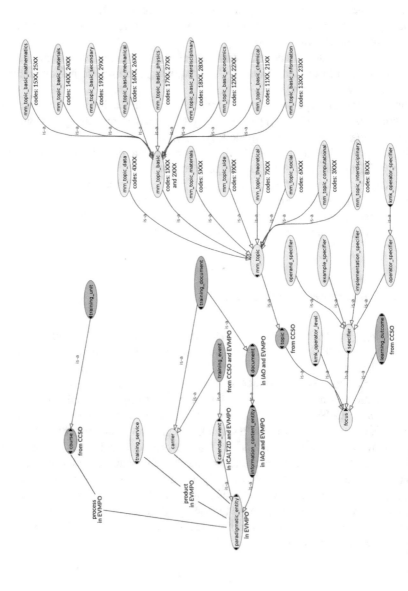

Fig. 3.7 Fragment of the OTRAS class hierarchy, version 1.0.5. The OWLViz protégé plugin was used to visualize the ontology [10]; solid lines and arrows labelled "is-a" denote subsumption ($\sqsubseteq$). Here, ICALTZD refers to the iCal ontology with time zones as datatypes [21]

"Upon successfully completing σ_1, participants can σ_2

with respect to σ_3

by doing σ_4;

for example, σ_5. (3.9)

Therein, σ_1 is the course or training material (carrier) for which a learning outcome is stated. The individuals $\sigma_2, \sigma_3, \sigma_4$ and σ_5 are specifiers (otras:specifier) of the learning outcome:

- σ_2 specifies the *operator* of the learning outcome (concept otras:operator_-specifier); a catalogue of operators with three-digit operator codes is included.
- σ_3 specifies the *operand* of the learning outcome (concept otras:operand_specifier); the operand can be formulated in terms of one or multiple topics, for which four-digit topic codes are given in OTRAS.
- σ_4 specifies the *implementation* (concept otras:implementation_specifier), describing how the competency is carried out in practice (e.g. "by writing C++ codes" or "by carrying out appropriate series of MD simulations"); this specifier is optional.
- σ_5 specifies an *example* (concept otras:example_specifier), illustrating how the competency might be applied to a particular special case (e.g. "if asked to develop a molecular model for caffeine, one might choose to parameterize a rigid coarse-grained model consisting of multiple Mie interaction sites"). This specifier is also optional.

The operator specifier σ_2 indicates what sort of activity is enabled by possessing a certain competency. Learning outcomes in course syllabi are typically formulated concisely, e.g. "the students will be able to apply statistical mechanics to problems from fluid phase thermodynamics." In this example, the operator *is expressed by* the predicate "apply." In the interest of the legibility of a syllabus (and the work involved in writing it), a precise definition of the meaning of the operator specifier is usually not provided, and the interpretation is left to the intuition of the reader.

In the interest of platform and institutional interoperability between training providers, it can nonetheless be helpful to reach an agreement on a more precise specification of the semantics associated with a learning outcome formulation. This has aspects of both semantic and pragmatic interoperabilities, such as where multiple instructors are expected to abide by the same syllabus and/or conduct exams that confirm the success of the learning effort at a specified level. It may also support the automated rendition of a syllabus in multiple languages. For this task, OTRAS relies on a catalogue of operators disseminated by the German *Kultusministerkonferenz* (KMK), facilitating the specification of learning outcomes in the natural sciences in a consistent way [36]. In OTRAS, each operator has a three-digit topic code (e.g. 235) and otras:is_expressed_by a concise predicate (e.g. "apply"), while otras:is_defined_by gives a more detailed explanation of its meaning; in the given case, "use a known idea, equation, principle, theory or law in a new situation" [36]. The KMK operators roughly correspond to elementary (operator codes 1xx), intermediate (2xx) and advanced (3xx) levels of learning; they are complemented here by additional operator individuals (operator codes 4xx) that are expected to be more adequate for expressing certain competencies that are typically attributed to expert personnel.

The operand specifier σ_3 may be taken from a taxonomy of materials modelling topics (subclasses of otras:mm_topic) that are included in OTRAS. The first hierarchy level (and part of the second level) of this taxonomy is shown in Fig. 3.7. This is widely used within VIMMP beyond the specification of competencies, e.g. to sort and retrieve documents by identifying the addressed topics and to indicate relevant areas of interest and fields of knowledge to be used to matchmaking by the *translation router* [14] app of the VIMMP marketplace platform. OTRAS also permits the specification of topics via CCS, a taxonomy developed by the Association for Computing Machinery [37], and PhySH, developed by the American Physical Society [38]. Syllabi can be associated with learning outcomes by means of the relation otras:aims_to $\sqsubseteq$ ccso:aimsToLO. If a competency is asserted as such, irrespective of how it has been acquired, σ_1 can be absent, in particular, wherever the relation vico:has_competency from VICO, see below, is used to characterize the background of an evmpo:expert.

The *VIMMP Communication Ontology* (VICO) covers metadata on messages exchanged at the digital marketplace platform and participants that interact at an EVMF-interoperable platform [14]. Through the LCC ontology, VICO incorporates the ISO 3166 standard for referring to countries and regions [39]. Types of interlocutors (subclasses of vico:interlocutor) are referred to—in accordance with the usual EMMC nomenclature—as consultants, data providers, end users, manufacturers, model providers, software owners, training providers, translators and guests.[13] The concept vico:interlocutor_group is instantiated by individuals associated with each of these groups, e.g. vico:software_owner individuals belong to the vico:interlocutor_group individual vico:IG_SOFTWARE_OWNER. The communication branch of the class hierarchy is visualized in Fig. 3.8.

Selected concepts from OTRAS and VICO:

- vico:academic_title: a titular rank that corresponds to an academic degree.
- otras:focus: a studied object, topic, training objective or an aspect or constitutive part thereof. Subclasses include otras:learning_outcome, otras:operator_level, otras:specifier and otras:topic.
- vico:interlocutor_tag: a descriptor that specifies properties of an interlocutor which may co-determine ability/suitability for trading with certain partners at a digital marketplace—indicating the country of residence/registration, whether the described interlocutor is engaged in military or nuclear research, etc.
- vico:message: a stand-alone communication (rather than an appendix).
- otras:mm_topic (materials modelling topic): a topic related to the subject area of materials modelling, understood broadly.

---

[13]VICO: https://purl.vimmp.eu/semantics/vico/vico.ttl (non-resolvable IRI), mirrored at http://www.molmod.info/semantics/vico.ttl (resolvable URL).

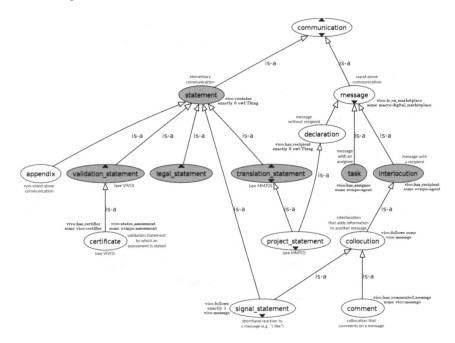

Fig. 3.8 Communication branch fragment of the VICO class hierarchy, version 1.2.6, including related EVMPO concepts. The OWLViz protégé plugin was used to visualize the ontology [10]; arrows labelled "is-a" denote subsumption (⊑)

- vico:person: a stand-alone agent that does not have multiple constituent parts or components each of which could act at a digital marketplace by themselves.
- otras:specifier: a constitutive element of a learning outcome (competency) description.
- otras:training_service: a tradeable object (evmpo:tradeable_object) that provides training contents or activities.
- otras:training_unit: an elementary (part of a) course that is not further subdivided into any smaller parts.

Selected relations from OTRAS and VICO:

- vico:contains; (:C contains :D) ⟺ :D is a proper part of :C, where :C and :D are both communications. Domain: evmpo:communication; range: evmpo:communication.
- vico:follows; (:C follows :D) ⟺ :C and :D are messages, and :C addresses or refers to :D. Domain: vico:message; range: vico:message.
- vico:has_affiliation indicates an institutional affiliation. Domain: vico:person; range: evmpo:institution.
- vico:has_author points to an agent that has issued a communication. Domain: evmpo:communication; range: evmpo:agent.

- otras:has_offered_course points to a course that is offered as part of the activities carried out as a training service. Domain: otras:training_service; range: otras:course.
- otras:has_specifier points to an operator, operand, implementation or example specifier of a learning outcome. Domain: otras:learning_outcome; range: otras:specifier.
- vico:is_certifier_of points to a certificate for which a certifier is (co-)responsible, having either issued the certificate or formally approved of its content. Domain: vico:certifier; range: vivo:certificate.
- otras:is_narrower_than; (:A is narrower than :B) $\Longleftrightarrow$:A and :B are topics such that if :A is a sign for an object, :B is also a sign for that object. This relation is defined to be a subproperty of skos:broader, cf. Isaac and Summers [40]; n.b., in SKOS, the relation is defined the other way around, i.e. (:A broader :B) $\Longleftrightarrow$:B is broader than :A. Domain: otras:topic; range: otras:topic. Analogous: otras:is_broader_than $\sqsubseteq$ skos:narrower.
- otras:is_part_of_course points to the course to which the given training unit belongs. Domain: otras:training_unit; range: otras:course.

References

1. F. Baader, I. Horrocks, C. Lutz, U. Sattler, *An Introduction to Description Logic* (Cambridge University Press, Cambridge, 2017)
2. T. Schneider, M. Šimkus, Ontologies and data management: a brief survey. Künstl. Intell. **34**(3), 329–353 (2020). https://doi.org/10.1007/s13218-020-00686-3
3. F. Berto, M. Plebani, *Ontology and Metaontology* (Bloomsbury, London, 2015)
4. D. Allemang, J. Hendler, *Semantic Web for the Working Ontologist*, 2nd edn. (Morgan Kaufmann, Waltham, 2011)
5. D. Fensel, T. Horrocks, F. Van Harmelen, S. Decker, M. Erdmann, M. Klein, OIL in a nutshell, in *Proceedings of EKAW 2000*, LNAI, vol. 1937 (Springer, Heidelberg, Germany, 2000), pp. 1–16
6. R. Fikes, D. McGuinness, An axiomatic semantics for RDF, RDF-S, and DAML+OIL. Note, W3C, Cambridge, Massachusetts, USA, 2001. https://www.w3.org/TR/daml+oil-axioms/
7. J. Hendler, D.L. McGuinness, The DARPA agent markup language. IEEE Intell. Syst. **15**(6), 67–73 (2001)
8. I. Horrocks, U. Sattler, Ontology reasoning in the SHOQ(D) description logic, in *Proceedings of IJCAI*, ed. by B. Nebel (Morgan Kaufmann, Los Altos, California, USA, 2001), pp. 199–204
9. C. Golbreich, M. Horridge, I. Horrocks, B. Motik, R. Shearer, *OBO and OWL: Leveraging Semantic Web Technologies for the Life Sciences*, LNCS, vol. 4825 (Springer, Heidelberg, Germany, 2007), pp. 169–182
10. M. Dudáš, S. Lohmann, V. Svátek, D. Pavlov, Ontology visualization methods and tools: a survey of the state of the art. Knowl. Eng. Rev. **33**, e10 (2018)
11. J.B. Lamy, *Python et les Ontologies* (ENI, Saint-Herblain, France, 2019)
12. B. Motik, B. Cuenca Grau, I. Horrocks, Z. Wu, A. Fokoue, C. Lutz, OWL 2 web ontology language profiles. W3C recommendation, W3C (Cambridge, Massachusetts, USA, 2012). https://www.w3.org/TR/2012/REC-owl2-profiles-20121211/
13. M.T. Horsch, S. Chiacchiera, M.A. Seaton, I.T. Todorov, D. Toti, G. Goldbeck, Introduction to the VIMMP ontologies. Technical report (2020). https://doi.org/10.5281/zenodo.3936796

14. M.T. Horsch, S. Chiacchiera, M.A. Seaton, I.T. Todorov, K. Šindelka, M. Lísal, B. Andreon, E.B. Kaiser, G. Mogni, G. Goldbeck, R. Kunze, G. Summer, A. Fiseni, H. Brüning, P. Schiffels, W.L. Cavalcanti, Ontologies for the Virtual Materials Marketplace. Künstl. Intell. **34**(3), 423–428 (2020). https://doi.org/10.1007/s13218-020-00648-9

15. G. Goldbeck, E. Ghedini, A. Hashibon, G.J. Schmitz, J. Friis, A reference language and ontology for materials modelling and interoperability, in *Proceedings of NWC*, NAFEMS (Knutsford, UK, 2019) p NWC_19_86 (2019)

16. EMMC Coordination and Support Action, European Materials and Modelling Ontology (2020), https://github.com/emmo-repo/, https://emmc.info/emmo-info/. Accessed 8 Apr 2020

17. A.F. De Baas (ed.), *What makes a material function?* (EU Publications Office, Luxembourg, Let me compute the ways, 2017)

18. EMMC Coordination and Support Action, EMMC Translation Case Template. Technical report (2017). Accessed 1 Mar 2020

19. D. Hristova-Bogaerds, P. Asinari, N.A. Konchakova, L. Bergamasco, A. Marcos Ramos, G. Goldbeck, D. Höche, O. Swang, G.J. Schmitz, EMMC Translators' Guide. Technical report, European Materials Modelling Council (2019). https://doi.org/10.5281/zenodo.3552260

20. CEN-CENELEC Management Centre, Materials modelling: terminology, classification and metadata. CEN workshop agreement 17284, Brussels, Belgium (2018)

21. D. Connolly, L. Miller, RDF Calendar: an application of the resource description framework to iCalendar data. Interest group note, W3C, Cambridge, Massachusetts, USA (2005), https://www.w3.org/TR/rdfcal/. Accessed 1 Jan 2020

22. W. Ceusters, An information artifact ontology perspective on data collections and associated representational artifacts. Stud. Health Technol. Inf. **180**, 68–72 (2012)

23. C.S. Peirce, *Peirce on Signs: Writings on Semiotic* (University of North Carolina Press, Chapel Hill, North Carolina, USA, 1991)

24. M.T. Horsch, S. Chiacchiera, B. Schembera, M.A. Seaton, I.T. Todorov, Semantic interoperability based on the European materials and modelling ontology and its ontological paradigm: mereosemiotics, in *Proceedings of WCCM-ECCOMAS 2020*, to appear (2021). https://doi.org/10.5281/zenodo.3902900

25. J. Ison, M. Kalaš, I. Jonassen, D. Bolser, M. Uludağ, H. McWilliam, J. Malone, R. López, S. Pettifer, P. Rice, EDAM: an ontology of bioinformatics operations, types of data and identifiers, topics and formats. Bioinform. **29**(10), 1225–1332 (2013)

26. N. Bassiliades, M. Symeonidis, P. Gouvas, E. Kontopoulos, G. Meditskos, I. Vlhavas, PaaSPort semantic model: an ontology for a platform-as-a-service semantically interoperable marketplace. Data Knowl. Eng. **113**, 81–115 (2018)

27. M.T. Horsch, C. Niethammer, G. Boccardo, P. Carbone, S. Chiacchiera, M. Chiricotto, J.D. Elliott, V. Lobaskin, P. Neumann, P. Schiffels, M.A. Seaton, I.T. Todorov, J. Vrabec, W.L. Cavalcanti, Semantic interoperability and characterization of data provenance in computational molecular engineering. J. Chem. Eng. Data **65**(3), 1313–1329 (2020)

28. M.T. Horsch, S. Chiacchiera, M.A. Seaton, I.T. Todorov, B. Schembera, P. Klein, N.A. Konchakova, Pragmatic interoperability and translation of industrial engineering problems into modelling and simulation solutions, in *Proceedings of DAMDID 2020*, to appear (2021). https://doi.org/10.5281/zenodo.3902873

29. G. Rutkai, J. Vrabec, Empirical fundamental equation of state for phosgene based on molecular simulation data. J. Chem. Eng. Data **60**(10), 2895–2905 (2015)

30. J.G. Barrientos, E.R. Cruz Sosa, P.E. García Castro, Considerations of e-commerce within a globalizing context. Int. J. Manag. Informat. Sys. **16**(1), 101–110 (2012)

31. K.H. Pries, J.M. Quigley, *Reducing Process Costs with Lean, Six Sigma, and Value Engineering Techniques* (CRC, Boca Raton, Florida, USA, 2013)

32. EMMC Coordination and Support Action, Report on business related quality attributes for industry integration of materials modelling. EMMC-CSA project deliverable 6.3, EU, Horizon 2020 research and innovation programme (2018)

33. J. Machač, F. Steiner, J. Tupa, Product life cycle risk management, in *Risk Management Treatise for Engineering Professionals*, ed. by C.F. Oudoza (London, UK, IntechOpen, 2018), pp. 51–72

34. X. Zhang, K. Li, C. Zhao, D. Pan, A survey on units ontologies: architecture, comparison and reuse. Prog. Electron. Lib. **51**(2), 193–213 (2017)
35. E. Katis, H. Kondylakis, G. Agathangelos, V. Kostas, Developing an ontology for curriculum & syllabus, in *Prof. ESWC Satellite Events*, ed. by A. Gangemi, A.L. Gentile, A.G. Nuzzolese, M.S. Rudolph, H. Maleshkova, J.Z. Paulheim, M. Alam Pan (Springer, Cham, Switzerland, 2018), pp. 55–59
36. Kultusministerkonferenz, Operatoren für die naturwissenschaftlichen Fächer (Physik, Biologie, Chemie) in englischer Sprache an den Deutschen Schulen im Ausland, version 265 (March 2014), technical report (KMK, Bonn,Germany, 2014)
37. Association for Computing Machinery, The 2012 ACM Computing Classification System. Technical report, ACM (2012). https://www.acm.org/publications/class-2012/
38. American Physical Society, Physics Subject Headings (2020). https://physh.aps.org/
39. J. Celko, *Data* (Measurements and Standards in SQL, Morgan Kaufmann, Burlington, Massachusetts, USA, 2010)
40. A. Isaac, E. Summers, SKOS Simple Knowledge Organization System primer. Working group note, W3C, Cambridge, Massachusetts, USA (2009). https://www.w3.org/TR/2009/NOTE-skos-primer-20090818/

Chapter 4
Semantic Technology for Simulations and Molecular Particle-Based Methods

4.1 Brief Overview of Ontologies for Modelling and Simulation

Since the appearance of ontologies in computer science in the 90s [1], there have been proposals and endeavours to use them to describe modelling and simulation, with the aim to support the exchange of information both between people (communication) and between software (interoperability) [2–4].

As with any topic, clearly also in this case many different points of view can be adopted. Browsing the literature on the subject, we identify two major perspectives: taking a more philosophical approach, some authors focus on the process of modelling itself, as a cognitive process, and its relation to the physical world, see, e.g. [3]; on the other side, with a more application-oriented view, other authors focus on the structuring of models and simulations, giving for granted their connection to reality [2, 4].

The first perspective might seem surprising, but, in fact, there are various intellectual steps that are undertaken each time we use a numerical simulation to make predictions about a certain real problem: typically these involve abstraction and simplification, to arrive to a model (in the MODA sense) and then its conversion into a numerical implementation (see, for example, [5–7], where the steps of the first part, what Robinson calls *conceptual modelling for simulation*, are described). It is therefore relevant to be able to formally describe these steps explicitly, for example, to compare models involving different levels of abstraction, and to address model verification and validation.

Coming to the ontologies presented or referred to in this book, EMMO and VIVO are close in spirit to the first perspective, whereas the ontologies we will describe in this chapter, the *VImmp Ontology of Software* (VISO) and the *Vimmp Ontology of Variables* (VOV), are closer in spirit to the second one. Before describing them (in Sects. 4.3 and 4.4), we highlight in the following some existing ontologies and assets that have a similar purpose or scope.

© The Author(s) 2021
M. Horsch et al., *Data Technology in Materials Modelling*,
SpringerBriefs in Applied Sciences and Technology,
https://doi.org/10.1007/978-3-030-68597-3_4

In the area of physics and engineering, we would like to point out two ontologies: the pioneering PhysSys [2], which already gave a central role to theories such as mereology and topology and recognized the need for different viewpoints on a given problem, and the very recent Physics-based Simulation Ontology (PSO) [4], which uses the Basic Formal Ontology (BFO) [8] as an upper ontology and is split into two parts addressing the physical phenomena (PSO-Phys) and the simulation aspects (PSO-Sim). Both ontologies focus on what in the EMMC vocabulary are called continuum models.

Looking at solutions to characterize software in other domains, we find that, in logistics and manufacturing, discrete-event simulations are the object of the DeMO Ontology [9]; recent work capturing the point of view of a scientist end user has led to the Software Ontology (SWO) [10] for life sciences and to OntoSoft [11] for geosciences.

Moving to variables, we would like to recall the catalogue for Quantities, Units, Dimensions and Data Types Ontologies (QUDT) [12], which addresses among others dimensional analysis and a classification of units, and the Scientific Variables Ontology (SVO) [13, 14]. The latter originated analysing thousands of variables in the area of natural sciences, but provides a framework that can be, in principle, adapted to other fields.

With a focus on the software engineering aspects, we highlight instead the Software Engineering Ontology Network (SEON) [15], an ontology network based on the Unified Foundational Ontology (UFO) [16, 17]. Connected to SEON, a Reference Ontology on Object-Oriented Code (OOC-O) was recently proposed [18].

We note that the relation of some of these ontologies to our work is very concrete: in fact, concepts from SWO and QUDT are currently imported in VISO, VOV and other VIMMP ontologies (cf. Sects. 4.3 and 4.4).

4.1.1   Examples of Applications

As already explained in Chap. 1, ontologies are an explicit and formal way to represent knowledge in a certain domain. But how are they *actually used* in the context of simulations and modelling?

This question connects to the purpose the ontology is designed for and to technical aspects, such as the availability and choice of tools (for example, to connect ontologies to programming languages[1]). And it also poses the question whether we expect the end users to be (mainly) humans or machines.

Also, the use could be more or less direct: thinking, for example, of a database, a triplestore would make an immediate use of the ontology, whereas a less direct approach would be to take into account aspects of the ontology when designing the database.

---

[1]For example, *Owlready 2* [19] is a Python module that allows to import and manipulate OWL 2.0 ontologies and do ontology-oriented programming in Python.

We can get some insight on the possibilities by looking at the examples given above: OntoSoft [11] was used to design a platform to find and compare software [20]; the Simulation Intent Ontology [21] is used in connection with the CoGui tool [22] to automatize some steps in the simulation setup; finally, one of the perspective uses of SVO and the connected tools [13, 14] is to generate suitably formed variables starting from free-form text.

From the point of view of the source code, a perspective use of OOC-O is to support polyglot programming [18], i.e. the simultaneous use of multiple object-oriented programming languages.

4.2 Other Relevant Assets and Approaches

In the previous section, we limited the scope to ontologies; however, as discussed in Chap. 1, the semantic spectrum is wide, and along with them there are other relevant assets, which are technically different but similar in spirit, such as *data schemas* (cf. Chap. 2).

Also, we should recall different branches of a field that is sometimes referred to as *conceptual modelling*. Historically, in the '60s–'70s, novel ideas setting the basis of this field appeared in different areas of computer science, namely, artificial intelligence, programming languages, databases, software engineering [23]: these ideas lead, among others, to the development of knowledge-representation languages, object-oriented programming and entity-relationship (ER) models (see [23] for a discussion of the pioneering ideas in each area and a brief history of the topic).

Even if the connections between these approaches are not always direct, the thinking behind their development is similar: so, for example, when building an ontology for a domain, it is definitely instructive to look also into object-oriented programs and schemas for such domain, and vice versa.

As an illustration of schemas for our area, we recall the Chemical Markup Language (CML) [24] and the ThermoML schema [25, 26], a IUPAC standard primarily developed at NIST. In the direction of object-oriented programs, a popular tool is the Atomic Simulation Environment (ASE) [27] which allows to set up, control, visualize and analyse simulations at the atomic and electronic level.

Another relevant topic is that of *visual programming*: a visual scheme is used to represent a model (in the general sense), but also to generate the source code (model-driven simulation). This operation can sometimes work also in the opposite direction, extracting the model from the source code, a form of reverse engineering.

Finally, in the area of software design and business modelling, it is important to recall the role of the Object Management Group (OMG) [28] that was formed 30 years ago; in particular, its activities lead to the development of the Unified Modeling Language (UML) [29] and an ecosystem of specifications based on it.

In a nutshell, UML allows to describe a system[2] structure and behaviour via different types of diagrams. It was motivated as a unifying object-oriented language, and one of its main aims is to "advance the state of the industry by enabling object visual modeling tool interoperability" [29]. The visual aspect of the diagrams can be used to share ideas; however, these diagrams can also be given "life", in the sense of visual programming.

The OMG standards are widely adopted and there are many (commercial and not) tools based on UML that allow, for example, to generate executable code, check the model and generate test suites.

Of course, "modeling" in the case of UML has a more general meaning (as abstraction) than as we intend it in the EMMC sense (as applied to materials and based on physics and mathematics). From the literature, we note that UML does not appear to be strongly connected to science applications, but it is probably used internally by professional software tools, including those commonly used in engineering.

Finally, an important contribution bridging between UML and ontologies is OntoUML, which is ontologically well-founded version of UML (more specifically, of the UML 2.0 fragment of class diagrams) [16, 17].

4.3 Software Capabilities

The aim of the *VImmp Ontology of Software* (VISO)[3] is to characterize software tools in the area of materials modelling, especially their features (i.e. capabilities), intended both at the model and solver level, but also their technical requirements, compatibility with other tools and licensing aspects. The concepts defined within this ontology will, first, guide the ingest of information on the VIMMP platform, and, later, allow the users to retrieve and compare tools. Below an upper level (viso-general, cf. Fig. 4.1) that addresses aspects common to all software, we split VISO into three branches focusing on classes of models: electronic (EL, viso-el), atomistic and mesoscopic (AM, viso-am), and continuum (CO, viso-co) models.[4] These branches depend on viso-general, but can be loaded independently of the other two siblings. We underline that both VISO and VOV (presented in the next section) are designed to address models from the four granularity levels of RoMM [30]. However, in this book, we

---

[2]A "system" is intended here in a very general sense, as something made of components.

[3]VISO: https://purl.vimmp.eu/semantics/viso/viso-general.ttl, https://purl.vimmp.eu/semantics/viso/viso-electronic.ttl, https://purl.vimmp.eu/semantics/viso/viso-atomistic-mesoscopic.ttl, https://purl.vimmp.eu/semantics/viso/viso-continuum.ttl (all of which are non-resolvable IRI); the concatenation of the four files is mirrored at http://www.molmod.info/semantics/viso-all-branches.ttl (resolvable URL).

[4]To avoid name clashes between the branches, prefixes are used as indicated. In the protégé editor, one can choose different options for the rendering (*view* tab), including rendering by short name and rendering by prefixed name.

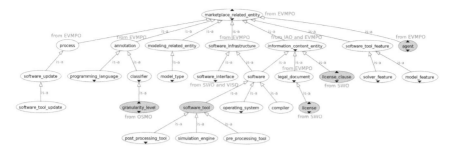

Fig. 4.1 Fragment of VISO showing its upper and intermediate classes and their connection to EVMPO and external assets; the diagram was generated using the OWLViz protégé plugin; grey arrows labelled "is-a" denote subsumption (⊑), i.e. rdfs:subClassOf

especially focus on its AM branch, which deals with molecular models and particle-based methods; for more details on the other branches, we refer the reader to [31] and to a recent VIMMP ontology release [32, 33].

Accordingly, selected major concepts from viso-general are:

- viso:software: a computer program. Its direct (mutually disjoint) subclasses are: viso:software_tool, viso:compiler, viso:operating_system.
- viso:programming_language: a language that can be used to write software.
- viso:software_tool_feature ≡ (viso:model_feature ⊔ viso:solver_feature): a capability of a software tool, intended as either a model aspect that can be addressed (viso:model_feature) or as a numerical algorithm which is implemented (viso:solver_feature). Following the approach from RoMM [30], these two classes are disjoint.
- viso:model_type: a classification of the model, intended as in RoMM [30].
- viso:model_object: the type of object entering the model and carrying degrees of freedom. Its subclasses in the AM branch (cf. Fig. 4.2) include viso-am:interaction_site, viso-am:interaction_surface, viso-am:connected_object.
- viso:software_update: it describes (as text) the changes between versions of a software. In particular, its subclass viso:software_tool_update allows to describe the addition/removal of features from a tool.
- viso:software_interface: an interface between a software and a user or a client (i.e. a program or device). Some subclasses of this class are taken from the SWO software interface class (swo:SWO_9000050) [10].
- viso:license: a regulation of the right to use, modify and distribute something, in this case software. It is declared to be equivalent to the Software Licence class from SWO (swo:SWO_0000002), cf. Malone et al. [10].
- viso:license_clause: it is equivalent to the Licence clause class from SWO (swo:SWO_9000005), cf. Malone et al. [10].

Selected relations (object properties) from VISO are:

- viso:has_feature points to a (model or solver) feature of a tool. Domain: viso:software_tool; range: viso:software_tool_feature.

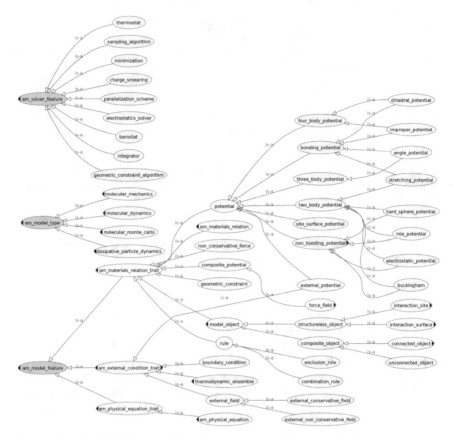

Fig. 4.2 Branch of VISO for atomistic-mesoscopic models (viso-am); the diagram was generated using the OWLViz protégé plugin; grey arrows labelled "is-a" denote subsumption (⊑), i.e. rdfs:subClassOf. A black (pointing left/right) triangle symbol within an ellipse indicates that some of the (super-/sub-) classes of the given one are omitted from the current visualization

- viso:is_compatible_with relates two tools that are able to exchange information directly, with no need to interface. Domain and range: viso:software_tool.
- viso:involves, i.e. (X involves Y) means that there is a mathematical expression or an algorithmic formulation of X that contains Y. Domain: viso:software_tool_ feature ⊔ viso:modeling_related_entity; range: vov:variable ⊔ vov:function ⊔ viso: model_object.
- viso:is_tool_for_model   associates   tools   with   models.   Domain: viso:software_tool; range: viso:model_type.
- viso:requires relates a tool to libraries and/or operating systems. Domain: viso:software_tool; range: viso:software.
- viso:is_modelling_twin_of relates two objects that (despite being possibly distinct individuals) are equivalent from the modelling point of view.

The upper part of VISO also contains a number of datatype properties, mostly Boolean ones (e.g. viso:is_free, viso:has_a_gpu_version), but not only (e.g. viso:has_main_name, viso:has_version_identifier) to characterize the software [32].

Below viso:general, the EL, AM and CO branches of VISO expand on the categories viso:model_feature, viso:solver_feature and viso:model_type (cf. Fig. 4.2 for the AM one). These classes are the richest ones of VISO, and they contain most of the concepts that are peculiar to our domain. The three branches have a common structure, in that the subclasses of viso:model_feature are further classified into (non-disjoint) classes of viso:materials_relation_trait, viso:physical_equation_trait and viso:external_condition_trait. For clarity, we systematically use *trait* here, and not *aspect*, since the latter keyword has a different and well-defined role within OSMO and MODA.

In the last part of this section, we look into more detail at the viso-am branch, which was designed considering Molecular Dynamics, Molecular Mechanics, Dissipative Particle Dynamics and Monte Carlo methods. First of all, our choice to treat together the atomistic and mesoscopic models is motivated by the fact that in many cases they rely on the same numerical methods and a given software tool can address both. Also, the meaning of "mesoscopic" within RoMM is different from the usual acceptation: as soon as two or more atoms are grouped into an entity, this is considered a mesoscopic model; since united-atom models already fall into this class, treating these two granularity levels jointly seems well justified.

It is important to underline that the RoMM [30] classification principle is based on what a modelling entity *represents* , a criterion that is indeed well suited to multiscale modelling. A complementary and quite natural classification could be based on the *mathematical nature* of the modelling entity: for example, the classical models could be distinguished in particle-based and field-based ones. While as a rule of thumb AM models are particle-based and CO models are field-based, typically there are also fields in AM models, discrete particles in CO and classical particles in EL ones. Above all, it is important to realize that the two classifications are fundamentally different: to give an extreme example, we could have a particle-based model of the solar system, where each particle represents a planet!

In this direction, an important concept in VISO is that of viso:model_object[5] (cf. Fig. 4.2) which is the type of object entering the model and carrying degrees of freedom. To be able to encompass different chemical objects, we need to adopt a neutral vocabulary; in the AM branch, we choose to use viso-am:interaction_site to indicate a point which is involved in (experiences) some interaction; it can represent the centre of a physical particle (an atom, a coarse-grained bead), but also be a fictitious particle. Similarly, a viso-am:connected_object, where connectedness is *via* bonds of some type, could be a molecule (in the chemical sense) or an aggregate. Finally, we have viso-am:interaction_surface, which is a surface affecting the interactions and is treated as continuum, not as a collection of sites; for example, it could be a wall. It is clear that our approach focuses on the mathematical nature of

---

[5]Not to be confused with the "Object model" concept of Chap. 2.

the objects, not on what they represent: this is a convenient point of view from the mathematical and numerical sides. Of course, to choose the appropriate Materials Relation we still need to know what the object represents.

Moving on to interactions, we highlight the classes viso-am:potential, viso-am:composite_potential and viso-am:non_conservative_force (cf. Fig. 4.2): the first one refers to the mathematical expression (functional form) of a potential energy and its elements are used as building blocks for elements of the second class; the second refers to a potential that is defined by more than just one single functional form acting between a pair of species; the last one refers to forces, typically appearing in coarse-grained models, that cannot be written in terms of a potential. Special cases of composite potentials are what in computational chemistry are known as Force Fields (called Interatomic Potentials in Physics): viso-am contains classes for some of the most popular ones [32].

So far, we have given examples that pertain to the materials relation, i.e. subclasses of viso-am:materials_relation_trait. Below viso-am:external_condition_trait, we find concepts as the boundary conditions, external fields and potentials, and the thermodynamic ensembles (cf. Fig. 4.2).

The class hierarchy for the solver features is much simpler than that for the model ones, being just a list of classes (including viso-am:integrator, viso-am:minimi-zation, ...), each populated by various individual algorithms.[6] We underline at this point that the splitting into solver and model features is not always straightforward, since it depends on how much relevance is given to an ingredient of the method: a prototypical example is that of thermostats, which are typically considered as purely numerical aspects, but have a central role for models such as Dissipative Particle Dynamics (cf. the discussion in [31]). To circumvent this problem and allow for different views while keeping solver and model features separated, we define in VISO the relation viso:is_modelling_twin_of.

Within VISO, we intentionally don't go beyond a certain level of detail in the description of software; in particular, the variables entering the models and algorithms are dealt with by the VOV ontology, presented in the next section.

4.4   Variables and Functions

The purpose of the *Vimmp Ontology of Variables* (VOV)[7] is to organize the variables (in a broad sense, including constants) that appear in modelling and simulations, and to connect them to models and algorithms in which they are *involved* and to model objects (e.g. entities entering a simulation, such as sites, rigid bodies) which they are *attached to*. VOV can be used in connection with VISO and OSMO to further specify models, algorithms and workflows. The main concepts from VOV are:

---

[6]We recall that individuals are not visible in the figures produced with OWLViz.

[7]VOV: https://purl.vimmp.eu/semantics/vov/vov.ttl (non-resolvable IRI), mirrored at http://www.molmod.info/semantics/vov.ttl (resolvable URL).

Fig. 4.3 Fragment of VOV showing selected subclasses of vov:variable and the subclasses of vov:function; the diagram was generated using the OWLViz protégé plugin; grey arrows labelled "is-a" denote subsumption (⊑), i.e. rdfs:subClassOf

- vov:variable: a variable in the mathematical sense, i.e. a symbol that stands for a quantity in a mathematical expression.
- vov:function: a relation between two or more variables (e.g. the radial distribution function, the energy density of states); it can be defined via a mathematical equation or via tabulated values. Its subclasses (cf. Fig. 4.3) include vov:field.

Variables in VOV can be classified according to three main criteria: by their scope (vov:object_variable, vov:pair_variable, vov:system_variable, vov:universal_variable), their rank (vov:scalar_variable, vov:vector_variable, vov:tensor_variable) or their basilar kind (vov:mass, vov:energy, …), for which qudt:QuantityKind is used [12]. In Fig. 4.3, we show the splitting of vov:variable according to scope and the subclasses

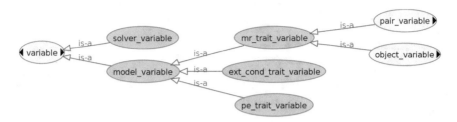

Fig. 4.4 Fragment of VOV showing selected subclasses of vov:variable, according to their usage; the diagram was generated using the OWLViz protégé plugin; grey arrows labelled "is-a" denote subsumption (⊑), i.e. rdfs:subClassOf

of vov:function. A different classification, also present in VOV and shown in Fig. 4.4, distinguishes variables based on the nature of the features they are involved in, as that is stated in VISO. Since a variable can be involved in multiple features of different nature, clearly the two main classes (vov:model_variable and vov:solver_variable) are not disjoint. Also, the vov:model_variable class is further split, mirroring in this way the hierarchy of model features in VISO.

Selected relations (object properties) from VOV are:

- vov:has_attached_variable points to a variable that is carried by/attached to an object. Its subproperties include vov:has_mass, vov:has_position and vov:has_velocity. Domain: viso:model_object; range: vov:object_variable.
- vov:has_attached_function points to a function that is carried by/attached to an object. Its subproperties include vov:has_velocity_field and vov:has_wavefunction. Domain: viso:model_object; range: vov:function.
- vov:shares_value_with indicates that two variables have the same numerical value. Domain and range: vov:variable.
- vov:shares_role_with indicates that two variables have the same role in the model or solver. This property can be used to instantiate variables that are already defined in VOV as individuals and to connect them to those. Domain and range: vov:variable.

Note that both relations specify viso:involves, i.e. vov:has_attached_variable ⊑ viso:involves and
vov:has_attached_function ⊑ viso:involves.

For properties such as the particle species (in a broad sense, chemical species in atomistic models, particle/object type in general), label and index, we use datatype properties (vov:has_species, vov:has_label, vov:has_index).

While for some typical variables it makes sense to define in the ontology named individuals, in other cases it is necessary to allow the user the freedom to define new ones. Accordingly, VOV provides different mechanisms to define the needed variables: one can directly use variables that are present in VOV as individuals (e.g. vov:TARGET_TEMPERATURE) or introduce customized ones populating VOV classes (e.g. defining elements of vov:object_mass). A third approach is to characterize the value and role of new variables using the relations vov:shares_value_with and vov:shares_role_with. The last method, while very convenient, has the drawback that it will not automatically transfer to the new variable all the properties of the prototype one, for example, its physical dimensions; however, this transfer can be taken care of when creating the data storage.

Coming to the class vov:function, which concerns relations between variables, in the case of classical particle-based models such a concept is mostly needed in the processing of data:[8] in Fig. 4.3, one can see classes such as vov:trajectory, vov:pair_distribution_function and vov:autocorrelation_function. The class vov:field, that has a more central role in continuum models, is however relevant also for particle-based ones: think, for example, of external spatially varying fields or of the density and velocity fields that are obtained processing the raw results.

To clarify how the two VIMMP ontologies we are discussing in this chapter are linked to each other, in Fig. 4.5, we highlight some classes from VISO and VOV, together with the main relations between them. In Fig. 4.6, we illustrate the same concepts with a more concrete example including individuals: the example is about a Molecular Dynamics (MD) software tool (we imagine it to be called "A_MD_TOOL") which has certain features (e.g. a velocity-Verlet integrator and a potential energy called "A_POTENTIAL") that in turn relate to variables (e.g. the simulation time step or the mass and velocity of an interaction site). These variables can extend to the whole system or be limited in scope to a model object. Considering the variable usage, the time step is a vov:solver_variable, whereas the site mass is a vov:model_variable. The idea behind the classification shown in Fig. 4.4 is to help to identify the variables that affect the physics of the system from those that do not,[9] or should not, and to recognize which part of the governing equations they enter. So, as soon as a variable is involved in some feature, we can infer which class it belongs to; however, since the classes are not disjoint, we *cannot exclude* it belongs to the sibling class too.

We note at this point the general and somewhat obvious, but practically relevant, fact that there is a delicate trade-off between the looseness of concept definitions and the ability to make informative inferences; and that is even more so given the assumptions under which ontologies by construction work[10] [34]. That is, to be able to make stringent inferences, we need to make explicit statements about class disjointness, individuals being different and so on. Otherwise, we can still extract information, but

---

[8]Following RoMM nomenclature, "processing" comprises any manipulation of the raw data obtaining solving the governing equations [30].

[9]In this respect, we note that even what looks like the most harmless of all concepts, the number of particles entering a simulation, poses already a classification problem: while for some finite systems the number of simulated particles has indeed a physical relevance and therefore should belong to the model; in other cases, as for infinite systems, it is more appropriately seen as a solver parameter, not different from the number of grid points used to solve differential equations. Within VOV, this and similar variables belong to the class vov:system_composition_variable and are not automatically classified as model/solver ones.

[10]In frameworks like the Semantic Web, where information is expected to be distributed across multiple and heterogeneous resources, two assumptions are typically made: one assumes that there could be more information out there, beyond the currently accessible one (Open World Assumption) and that individuals may be named differently in different contexts (Non-Unique Naming Assumption) [34].

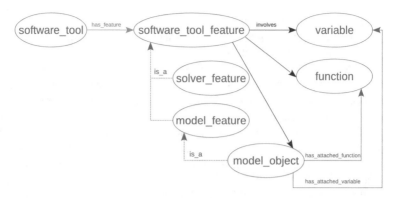

Fig. 4.5 Highlight of classes from VISO and VOV (shown as ellipses) and the relations between them (shown as arrows). The relations are: is_a, involves, has_attached_variable and has_attached_function

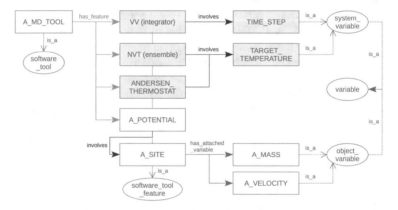

Fig. 4.6 An example containing individuals from VISO and VOV: full rectangles indicate individuals that are present in the ontologies, whereas empty rectangles indicate possible user-defined ones. Ellipses denote classes and arrows denote relations. The relations are: is_a (thin-head arrow for instantiation and thick one for class subsumption), involves and has_attached_variable

with the limitations just described. So far for inferences; nevertheless, of course, there is always the possibility to make explicit *negative statements* about variables and other concepts (using owl:complementOf, or owl:NegativePropertyAssertion, available in OWL2).

4.4.1 Simulation Variables vs Physical Properties

We open here a parenthesis to discuss the relation between the variables entering a simulation and the physical properties of real-world objects; this naturally leads to a comparison of simulations and experiments.

Let's consider, for example, a certain material (say, liquid water) and focus on its electric permittivity ϵ. The latter is a quantity that appears in electrostatic and electrodynamic laws (e.g. Coulomb's law), and can be measured experimentally based on them.

When designing a model for such material that, in particular, captures its electric permittivity, different approaches are possible: (a) we might take ϵ as fixed (a model parameter, a constant needed as an input), or (b) we could design a model containing dynamic degrees of freedom that carry electric dipoles, so that ϵ is an emergent property, and the input of the model is instead the properties of the degrees of freedom, possibly fictitious particles. In both cases, ϵ will be matched to the experimental value: in one case directly, and in the other by tuning the parameters associated with the simulated entities.

In the case (b), the value of ϵ can be estimated using liquid-state theory and computed from simulations using linear response theory: both procedures are quite far from what is done experimentally. However, a way to test that the model actually behaves as it should is to compute the reduction in the force between two fixed ions due to the presence of the medium; the same test can be done in the case (a), as a basic check of the numerical implementation of electrostatics, for example.

Now, is this so different from an experiment done on the material? Simulations, especially those involving some stochastic element, are in many ways similar to experiments, for example, they also require several repetitions and their results are affected by statistical errors.

In our view, the concept of *observation* (which includes *measurement*) in the EMMO could be generalized to accommodate also calculations (analytical and numerical ones), and the variables entering the models (both as input and output) that are numerical counterparts of physical properties could be recognized as such.

4.5 EngMeta and VIMMP Ontologies

The EngMeta scheme described in Chap. 2 and the VIMMP ontologies presented in this and in the previous chapter have a relevant overlap in scope, with similar keywords appearing in both assets. From Fig. 2.2, one can see that EngMeta includes concepts that on VIMMP side are addressed by different ontologies, in particular, OTRAS (e.g. author, publication, citation), VISO (e.g. software, force-field), VOV (e.g. variable), MMTO (e.g. project), OSMO (e.g. system component/material) and VICO (e.g. persons and organizations). The technical metadata (e.g. file size, checksum), instead, are mostly tackled by the Zontal storage itself [35, 36].

At the syntactic level, we are comparing an XSD schema and OWL DL ontologies: what are *entities* (say, "Software") and *attributes* (say, "license" and "softwareVersion") within EngMeta, typically correspond to *classes* within VIMMP ontologies (in this case, viso:software_tool) and to the objects or data a relation points to (in this case, a viso:license, pointed to by viso:has_license and a xs:string, pointed to by viso:has_version_identifier).

In general, to match or integrate two assets that, like here, differ syntactically and (even if slightly) semantically, one can think of performing the operation in two steps: a syntactic conversion first, then a semantic matching or integration (cf. Sect. 5.3).

4.6　Closing Thoughts

Clearly, there are many concepts involved here: formalization, standardization and automation. One could argue that for the domain we are interested in, i.e. simulations of materials, physics and mathematics *are* already universal languages: why, where and what kind of further formalization and standardization are needed?

For example, imagine we are given a set of equations that models the mixing of two fluids in an industrial device. Even if the mathematical formulation will be accessible to everybody with a scientific background, this does not capture the context the model is embedded in (in fact, the simulation intent, the assumptions and approximations made are normally expressed in natural language in an accompanying paper), and understanding it will require delving into a jungle of details. Also, importantly, the tacit assumptions and the technical jargon can vary a lot across communities (with the same algorithm having a different name and so on). Classifications and standardization can therefore help inter-community communication and collaboration and facilitate intra-community reuse of models. Coming to automation, of course, the possibility to generate source code from a pseudocode is very appealing, both for non-experts and for experts.

References

1. T.R. Gruber, Toward principles for the design of ontologies used for knowledge sharing? Int. J. Human-Comput. Stud. **43**(5), 907–928 (1995)
2. P. Borst, H. Akkermans, J. Top, Engineering ontologies. Int. J. Human-Comput. Stud. **46**(2–3), 365–406 (1997)
3. C. Turnitsa, J.J. Padilla, A. Tolk, Ontology for modeling and simulation, in *Proceedings of WSC*, ed. by B. Johansson, S. Jain, J. Montoya Torres (IEEE, Piscataway, New Jersey, USA, 2010), pp. 643–651
4. H. Cheong, A. Butscher, Physics-based simulation ontology: an ontology to support modelling and reuse of data for physics-based simulation. J. Eng. Des. 30(10–12, SI):655–687 (2019)
5. S. Robinson, Conceptual modelling for simulation Part I: definition and requirements. J. Oper. Res. Soc. **59**(3), 278–290 (2008)

6. S. Robinson, A tutorial on simulation conceptual modeling, in *Proceedings of WSC*, ed. by W.K.V. Chan, A. D'Ambrogio, G. Zacharewicz, N. Mustafee, G. Wainer, E. Page (IEEE, Piscataway, New Jersey, USA, 2017), pp. 565–579
7. S. Robinson, Conceptual modelling for simulation part II: a framework for conceptual modelling. J. Operat. Res. Soc. **59**(3), 291–304 (2008)
8. R. Arp, B. Smith, A.D. Spear, *Building Ontologies with Basic Formal Ontology* (MIT Press, Cambridge, Massachusetts, 2015)
9. G.A. Silver, J.A. Miller, M. Hybinette, G. Baramidze, W.S. York, DeMO: an ontology for discrete-event modeling and simulation. Simulation **87**(9), 747–773 (2011)
10. J. Malone, A. Brown, A.L. Lister, J. Ison, D. Hull, H. Parkinson, R. Stevens, The software ontology (SWO): a resource for reproducibility in biomedical data analysis, curation and digital preservation. J. Biomed. Semant. **5**, 25 (2014)
11. Y. Gil, V. Ratnaka, D. Garijo, OntoSoft: capturing scientific software metadata, in *Proceedings of K-CAP 2015*, ed. by K. Barker, J.M. Gómez Pérez (ACM, New York, USA, 2015), p. 32
12. X. Zhang, K. Li, C. Zhao, D. Pan, A survey on units ontologies: architecture, comparison and reuse. Prog. Electron Lib. **51**(2), 193–213 (2017)
13. M. Stoica, Scientific Variables Ontology (SVO) (2019), http://www.geoscienceontology.org/. Accessed 14 Jul 2020
14. M. Stoica, S. Peckham, Incorporating new concepts into the scientific variables ontology, in *Proceedings of eScience*, ed. by N. Williams (IEEE, Piscataway, New Jersey, USA, 2019), pp. 539–540
15. F. Borges Ruy, R. de Almeida Falbo, M. Perini Barcellos, S. Dornelas Costa, G. Guizzardi, SEON: a software engineering ontology networr, in *Proceedings of EKAW 2016*, ed. by E. Blomqvist, P. Ciancarini, F. Poggi, F. Vitali, LNCS, vol. 10024 (Springer, Cham, Switzerland, 2016), pp. 527–542
16. G. Guizzardi, Ontological foundations for structural conceptual models. Ph.D. thesis, University of Twente (2005)
17. G. Guizzardi, G. Wagner, J.P. Andrade Almeida, R.S.S. Guizzardi, Towards ontological foundations for conceptual modeling: The unified foundational ontology (UFO) story. Appl. Ontol. **10**(3–4), 259–271 (2015)
18. C.Z. de Aguiar, Rd.A. Falbo, V.E. Silva Souza, OOC-O: a reference ontology on object-oriented code, in *Proceedings of ER 2019*, ed. by A.H.F. Laender, B. Pernici, E.P. Lim, M. Palazzo, J. de Oliveira (Conceptual Modeling) LNCS, vol. 11788 (Springer, Cham, Switzerland), pp. 13–27
19. J.B. Lamy, *Python et les Ontologies* (ENI, Saint-Herblain, France, 2019)
20. OntoSoft Project, OntoSoft Portal (2020), https://www.ontosoft.org/portal/. Accessed 24 Mar 2020
21. F. Boussuge, C.M. Tierney, H. Vilmart, T.T. Robinson, C.G. Armstrong, D.C. Nolan, J.C. Leon, F. Ulliana, Capturing simulation intent in an ontology: CAD and CAE integration application. J. Eng. Des. **30**(10–12, SI), 688–725 (2019)
22. Laboratoire d'Informatique, de Robotique et de Microeléctronique de Montpellier (LIRMM) (2019) CoGui tool, https://www.lirmm.fr/cogui/. Accessed 24 Mar 2020
23. J. Mylopoulos, Information modeling in the time of the revolution. Inf. Syst. **23**(3–4), 127–155 (1998)
24. P. Murray-Rust, H.S. Rzepa, M. Wright, Development of chemical markup language (CML) as a system for handling complex chemical content. New J. Chem. **25**(4), 618–634 (2001)
25. M. Frenkel, R.D. Chirico, V. Diky, Q. Dong, K.N. Marsh, J.H. Dymond, W.A. Wakeham, S.E. Stein, E. Königsberger, A.R.H. Goodwin, XML-based IUPAC standard for experimental, predicted, and critically evaluated thermodynamic property data storage and capture (ThermoML). Pure Appl. Chem. **78**(3), 541–612 (2006)
26. M. Frenkel, R.D. Chirico, V. Diky, P.L. Brown, J.H. Dymond, R.N. Goldberg, A.R.H. Goodwin, H. Heerklotz, E. Königsberger, J.E. Ladbury, K.N. Marsh, D.P. Remeta, S.E. Stein, W.A. Wakeham, P.A. Williams, Extension of ThermoML: the IUPAC standard for thermodynamic data communications (IUPAC recommendations 2011). Pure Appl. Chem. **83**(10), 1935–1967 (2011)

27. A.H. Larsen, J.J. Mortensen, J. Blomqvist, I.E. Castelli, R. Christensen, M. Dułak, J. Friis, M.N. Groves, B. Hammer, C. Hargus, E.D. Hermes, P.C. Jennings, P.B. Jensen, J. Kermode, J.R. Kitchin, E.L. Kolsbjerg, J. Kubal, K. Kaasbjerg, S. Lysgaard, J. Bergmann Maronsson, T. Maxson, T. Olsen, L. Pastewka, A. Peterson, C. Rostgaard, J. Schiøtz, O. Schütt, M. Strange, K.S. Thygesen, T. Vegge, L. Vilhelmsen, M. Walter, Z. Zeng, K.W. Jacobsen, The atomic simulation environment: a Python library for working with atoms. J. Phys. Cond. Math. **29**, 273002 (2017)
28. Object Management Group (OMG) (2020), https://www.omg.org/. Accessed 24 Mar 2020
29. Object Management Group (OMG), Unified Modeling Language (UML) 2.0: Formal specification, version 2.5.1 (2017), https://www.omg.org/spec/UML/2.5.1/PDF. Accessed 27 Mar 2020
30. A.F. de Baas (ed.), *What Makes a Material Function?* (EU Publications Office, Luxembourg, Let me compute the ways, 2017)
31. M.T. Horsch, C. Niethammer, G. Boccardo, P. Carbone, S. Chiacchiera, M. Chiricotto, J.D. Elliott, V. Lobaskin, P. Neumann, P. Schiffels, M.A. Seaton, I.T. Todorov, J. Vrabec, W.L. Cavalcanti, Semantic interoperability and characterization of data provenance in computational molecular engineering. J. Chem. Eng. Data **65**(3), 1313–1329 (2020)
32. VIMMP Consortium, VIMMP ontologies: release dated 6th July 2020 (2020), https://www.vimmp.eu/?p=349. Accessed 7 Sep 2020
33. M.T. Horsch, S. Chiacchiera, M.A. Seaton, I.T. Todorov, D. Toti, G. Goldbeck, Introduction to the VIMMP ontologies. Technical report (2020). https://doi.org/10.5281/zenodo.3936796
34. D. Allemang, J. Hendler, *Semantic Web for the Working Ontologist*, 2nd edn. (Morgan Kaufmann, Waltham, Massachusetts, 2011)
35. D. Della Corte, W. Colsman, B. Welker, B. Rennick, Library eArchiving with ZONTAL space and the allotrope data format. Digital Libr. Perspect. **36**, 69–77 (2020). https://doi.org/10.1108/DLP-09-2019-0036
36. M.T. Horsch, S. Chiacchiera, M.A. Seaton, I.T. Todorov, K. Šindelka, M. Lísal, B. Andreon, E.B. Kaiser, G. Mogni, G. Goldbeck, R. Kunze, G. Summer, A. Fiseni, H. Brüning, P. Schiffels, W.L. Cavalcanti, Ontologies for the Virtual Materials Marketplace. Künstl. Intell. (2020). https://doi.org/10.1007/s13218-020-00648-9

Chapter 5
Applications of the Metadata Standards

5.1 Representing Scenarios

The division of a knowledge base $\mathcal{K} = (\mathcal{T}, \mathcal{A})$ into an ontology $\mathcal{T}$ and a scenario $\mathcal{A}$, as introduced in Sect. 3.1, is not only formal, but also motivated by practice. Fulfilling the role of a schema, an ontology needs to be ingested into a data infrastructure, or implemented by it, only once; frequent updates are undesirable, since they require a reannotation of data. For the scenarios handled by digital platforms, obversely, data retrieval and ingest are routine operations, and so are updates, since they need to occur whenever the represented reality changes, e.g. a new service is offered or a new user is registered. Challenges related to I/O (or ingest and retrieval) mainly concern the scenarios, not the ontologies, and their standardized representation by files, streams or protocols is the main vehicle for syntactic interoperability.

Since the IRIs of resources on the semantic web can point to each other as freely as the URLs of sites on the World Wide Web, i.e. in a graph-like way, it is natural to visualize scenarios by graphs. These representations are referred to as knowledge graphs. In Sect. 3.1, a scenario was defined as a tuple $\mathcal{A} = (\mathbf{I}, A_c, A_r, H)$ with individual names $\mathbf{I}$, conceptual assertions A_c, relational assertions A_r and elementary datatype property assertions H. The corresponding knowledge graph is a labelled graph $G = (\mathbf{I}, E, \Lambda_v, \Lambda_e)$ where the vertices are given by $\mathbf{I}$ and the edges by

$$E = \{(I, J) \mid \exists R \in \mathbf{R} : (I, R, J) \in A_r\} \subseteq \mathbf{I}^2. \tag{5.1}$$

Vertices are labelled according to the function $\Lambda_v : \mathbf{I} \to 2^{\mathbf{C} \cup \mathbb{R} \cup \Sigma^\star}$ that maps[1] each individual name $I \in \mathbf{I}$ to a set of labels

$$\Lambda_v(I) = (A_c(I) \cap \mathbf{C}) \cup \{v \in \Sigma^\star \mid \exists k \in \Sigma^\star : (k, v) \in H(I)\}, \tag{5.2}$$

---

[1] Notation: $2^{\mathbf{C} \cup \mathbb{R} \cup \Sigma^\star}$ is the power set (i.e. set of sets) over concept names for labelling individuals by class, reals for numerical datatype properties (including Booleans with 1 for true, 0 for false, as explicitly permitted by XSD) and words for textual datatype properties.

© The Author(s) 2021
M. Horsch et al., *Data Technology in Materials Modelling*,
SpringerBriefs in Applied Sciences and Technology,
https://doi.org/10.1007/978-3-030-68597-3_5

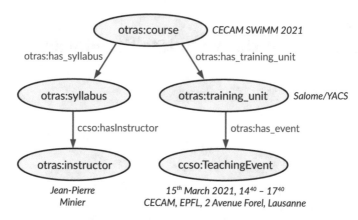

Fig. 5.1 Part of the knowledge graph corresponding to a scenario describing a training event, using the ontology OTRAS in combination with the Course Curriculum and Syllabus Ontology (CCSO) [2], cf. Sect. 3.5. The elliptical vertices represent individuals, labels inside the ellipses denote the concept instantiated by the respective individuals and labels in italics represent values associated with the individuals by means of elementary datatype properties, while the arrows (edges) represent relations between individuals and are labelled with the respective relation names

while the edge labelling function $\Lambda_e : E \rightarrow 2^{\mathbf{R}}$ assigns the corresponding relation names

$$\Lambda_e((I, J)) = \{R \in \mathbf{R} \mid (I, R, J) \in A_r\} \tag{5.3}$$

to an edge $(I, J) \in E$.

An example is given in Fig. 5.1; this knowledge graph might be read as follows: "There is a course labelled 'CECAM SWiMM 2021'. This course has a syllabus, in which information is given on an instructor who is labelled 'Jean-Pierre' and 'Minier'. The course has a training unit labelled 'Salome/YACS' for which event information is given," etc. While this particular representation does not contain IRIs of datatype properties (to match the definition of the knowledge graph given above), it could easily be modified to incorporate this information as well, e.g. by using property graphs following Abad Navarro et al. [1]. The individual name IRIs are not shown in the figure to simplify the visualization; however, they are included in the definition of the knowledge graph.

The technical implementation of semantic interoperability requires a syntactic representation by which information can be extracted from (or ingested into) a digital platform including a knowledge base; cf. Fig. 5.2 for a typical multi-tier design approach. For this purpose, subject-predicate-object triples can be employed, e.g. in TTL format (cf. Sect. 3.1), by which the scenario from Fig. 5.1 is rendered as follows:

```
@prefix ccso:  <https://w3id.org/ccso/ccso#>.
@prefix ical:  <https://www.w3.org/2002/12/cal/icaltzd#>.
@prefix otras: <https://purl.vimmp.eu/semantics/otras/otras.ttl#>.
@prefix swimm: <https://purl.vimmp.eu/semantics/example/swimm.ttl#>.
```

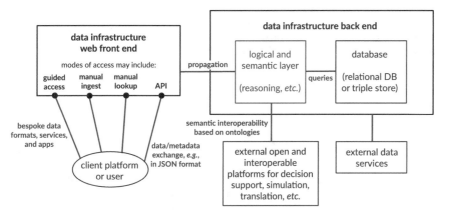

Fig. 5.2 Role of semantic technology within interoperable data infrastructures, illustrated for the case of a typical multi-tier architecture; a multitude of such platforms, which may be substantially more complex than outlined here, has been emerging in recent years. JSON is often used as a convenient format for communicating object data through HTTP-based APIs. Ontologies support reasoning in the logical application layer as well as interoperability between multiple platforms

```
@prefix vico: <https://purl.vimmp.eu/semantics/vico/vico.ttl#>.
@prefix xs: <http://www.w3.org/2001/XMLSchema#>.

swimm:SUMMER_SCHOOL a otras:course;
    ccso:csName "CECAM SWiMM 2021"^^xs:string;
    otras:has_syllabus swimm:COURSE_SYLLABUS;
    otras:has_training_unit swimm:TRAINING_UNIT_1.
swimm:COURSE_SYLLABUS a otras:syllabus;
    ccso:hasInstructor swimm:INSTRUCTOR.
swimm:TRAINING_UNIT_1 a otras:training_unit;
    otras:has_event swimm:EVENT_1;
    ccso:csName "Salome/YACS"^^xs:string.
swimm:EVENT_1 a ccso:TeachingEvent;
    ical:location "CECAM, EPFL, 2 Avenue Forel, Lausanne"^^xs:string;
    ical:dtstart "2020-03-15T14:40:00+01:00"^^xs:dateTime;
    ical:dtend "2020-03-15T17:40:00+01:00"^^xs:dateTime.
swimm:INSTRUCTOR a otras:instructor;
    vico:has_first_name "Jean-Pierre"^^xs:string;
    vico:has_last_name "Minier"^^xs:string.
```

Above, @prefix statements introduce the abbreviations employed for IRI prefixes, e.g. the datatype property https://w3id.org/ccso/ccso#csName is abbreviated by ccso:csName. The elementary datatypes follow the conventions for XML schemas, cf. Chapter 2.

TTL notation has the advantage that it can be employed consistently for the whole knowledge base, including both the ontology and the scenario. For many applications, however, this is more problematic than beneficial, because the expressive power of OWL and its various serializations (including TTL) goes far beyond what is needed to represent objects and their properties; consequently, it is harder to parse and to process. Moreover, it cannot be ensured at the syntax level that only information on the scenario is included. Instead, JavaScript Object Notation (JSON) is often preferred, particularly in its JSON Linked Data (JSON-LD) variety which was specifically designed for the purpose of exchanging semantically characterized information on objects and their relations. In JSON-LD format, the example scenario becomes

```
{
    "@context": {
        "ccso": "https://w3id.org/ccso/ccso#",
        "ical": "https://www.w3.org/2002/12/cal/icaltzd#",
        "otras": "https://purl.vimmp.eu/semantics/otras/otras.ttl#",
        "swimm": "https://purl.vimmp.eu/semantics/example/swimm.ttl#",
        "vico": "https://purl.vimmp.eu/semantics/vico/vico.ttl#",
        "xs": "http://www.w3.org/2001/XMLSchema#"
    },
    "@id": "swimm:SUMMER_SCHOOL",
    "@type": "otras:course",
    "ccso:csName": "CECAM SWiMM 2021",
    "otras:has_syllabus": {
        "@id": "swimm:COURSE_SYLLABUS",
        "@type": "otras:syllabus",
        "ccso:hasInstructor": {
            "@id": "swimm:INSTRUCTOR",
            "@type": "otras:instructor",
            "vico:has_first_name": "Jean-Pierre",
            "vico:has_last_name": "Minier"
        }
    },
    "otras:has_training_unit": {
        "@id": "swimm:TRAINING_UNIT_1",
        "@type": "otras:training_unit",
        "ccso:csName": "Salome/YACS",
        "otras:has_event": {
            "@id": "swimm:EVENT_1",
            "@type": "ccso:TeachingEvent",
            "ical:location": "CECAM, EPFL, 2 Avenue Forel, Lausanne",
            "ical:dtstart": "2020-03-15T14:40:00+01:00",
            "ical:dtend": "2020-03-15T17:40:00+01:00"
        }
    }
}
```

There, every pair of curly braces encloses the description of an object (except the value of @context, which includes the IRI prefix definitions), given as a sequence of key-value pairs. The individual names are provided as values corresponding to the key @id, while the instantiated concept names are indicated by the key @type. The other keys are relation names, and the associated values are the third elements of the respective triples, as can be seen from the direct correspondence between the TTL and JSON-LD examples given above.

Additionally, domain-specific solutions on the basis of the hierarchical data format HDF5 facilitate combining a greater volume of data, including binary data, with the corresponding semantic annotation [3], e.g. the H5MD format [4] for semantically enriched data in molecular modelling and simulation. The VIMMP marketplace platform API and its Zontal Space back end permit handling annotated digital objects through the HDF5-based Allotrope Data Format (ADF) [5–7].

5.2 Top-Level Ontology

For a fundamental philosophical underpinning, the European Materials and Modelling Ontology [8, 9] relies on a combination of physicalist mereotopology following Varzi [10] and a nominalist reinterpretation of Peirce's semiotics [11]. Therein, physicalist mereotopology primarily addresses the description of materials, which is extended by nominalist semiotics to describe modelling, simulation and experiments. For a discussion of nominalism, cf. Lewis [12], more specific implications of the approach of the EMMO on representing modelling and simulation of physical systems have been discussed elsewhere [13].

To facilitate the top-level ontology alignment of the VIMMP ontologies, a module with a scaled-down EMMO in TTL format is included, *EMMO version 1 simplified* (EMMO1s), which at the present stage (version 1.0.4) is based on EMMO version 1.0.0 alpha 2 (April 2020). EMMO1s provides user-friendly IRIs for EMMO concepts,[2] retaining the labels, e.g. the IRI of the EMMO concept with rdfs:label "Semiosis" is given in the original EMMO as emmo-semiotics:EMMO_008fd3b2_ 4013_451f_8827_52bceab11841. For these entities, EMMO1s specifies aliases that can be accessed directly through the label, such as emmo1s:Semiosis. In the interest of notational clarity, to indicate the origin of the concept definitions and the respective EMMO modules, these entities will here be denoted by the EMMO prefix followed by the EMMO1s suffix, e.g. by emmo-semiotics:Semiosis, even though internally, for VIMMP, it is actually emmo1s:Semiosis.

The *VIMMP Primitives* (VIPRS) module amplifies the ways in which the EMMO-based top-level semantic interoperability architecture can be applied to the relations characterizing metadata from the VIMMP marketplace-level domain ontologies.[3] With this aim, VIPRS extends the EMMO system of top-level relations by three features:

1. modal logic (e.g. Kripke semantics) and modal squares of opposition;
2. concatenation of mereotopological and semiotic relations, yielding mereosemiotic relations;
3. top-level datatype properties.

---

[2]EMMO1s: `https://purl.vimmp.eu/semantics/alignment/emmo1s.ttl` (non-resolvable IRI), mirrored at `http://www.molmod.info/semantics/emmo1s.ttl` (resolvable URL).

[3]VIPRS: `https://purl.vimmp.eu/semantics/alignment/viprs.ttl` (non-resolvable IRI), mirrored at `http://www.molmod.info/semantics/viprs.ttl` (resolvable URL).

While the EMMO can be used to describe materials and models as such, statements on necessity and possibility anchored in modal logic are metaontological, i.e. beyond the ontology, from the point of view of the EMMO [9], e.g. within the framework of the EMMO, an event can be described as a physical process, but the statement that "this process can possibly occur, but it will not necessarily occur" cannot be expressed. The present domain ontologies, however, make ample use of relations that are ultimately modal to specify capabilities (it *is possible* that :X will be used to do :Y) or requirements (if *is necessary* that if :X occurs, :Y also occurs).

To provide a top-level structure for modal relations, VIPRS includes *modal squares of opposition*,[4] cf. Fig. 5.3, by which the presence of individuals in a knowledge base can be associated with statements on whether their occurrence is possible, necessary, factual or fictional [16]. The modal operators can be given a variety of interpretations, depending on the precise use that is made of the ideas of necessity ($\square$) and possibility ($\lozenge$), respectively [17]; similarly, the definition of "occurrence" depends on the use that is made of the ontology and may depend on context—VIPRS accepts this ambiguity in order to be applicable to diverse types of knowledge bases and infrastructures. The term "to occur" in $\lozenge(\text{occ}[:X])$, ":X may occur," and similar, is employed to refer to the (possible or necessary) appearance of an individual :X in a certain type of environment, e.g. as an element of a valid simulation workflow. On this basis, relations concerning the possible or necessary co-occurrence of multiple individuals are defined, e.g. viprs:n_loc_or_rnoc (and others following the same pattern, cf. Fig. 5.3), where the IRI is to be read as "necessarily, the left occurs or the right does not occur"

$$I \text{ viprs:n\_loc\_or\_rnoc } J \equiv \square(\text{occ}[I] \vee \neg\text{occ}[J]) \equiv \square(\text{occ}[J] \rightarrow \text{occ}[I]).$$
$$(5.4)$$

cf. Fig. 5.3. Thereby, "occurrence" (by appearing in a certain type of environment) is not the same as "existence," i.e. presence in a knowledge base. It is in this sense that VIPRS can be employed as an implementation of possible-world semantics, Kripke semantics and/or ontological Meinongianism [16], even though it does not necessarily presuppose the use of any of these paradigms. The conceptualization relation

$$C \sqsubseteq S \sqcap \text{viprs:n\_loc\_or\_rnoc},$$
$$(5.5)$$

with $K_1 \, C \, I$ to be read as "K_1 conceptualizes I," relates a more (or equally[5]) generic individual to a more (or equally) specific one; it is used to introduce a step of abstraction into the modal co-occurrence relations, e.g. "necessarily, the left occurs *conceptual-or* the right does not occur"

---

[4]A square of opposition, going back to Aristotle, is a diagram containing four related statements, concepts or predicates labelled **A** for universal affirmation ("all"), **E** for universal negation ("no"), **I** for existential affirmation ("some") and **O** for existential negation ("not all"); cf. Westerståhl [14, 15] for a variety of applications.

[5]C is reflexive, $\forall K \in \mathbf{I}: \, K \, C \, K$.

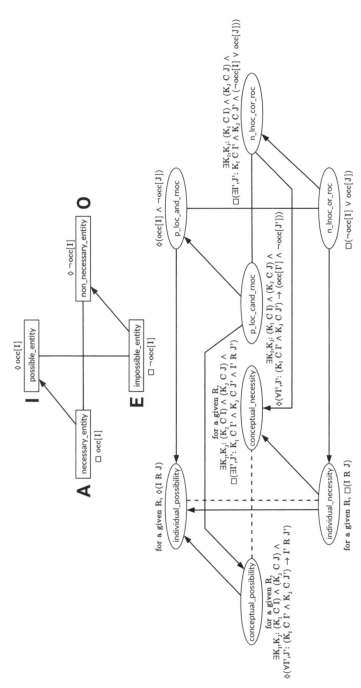

Fig. 5.3 Selected traditional (top) and generalized (bottom) modal squares of opposition from VIPRS. Here, occ[I] stands for "I occurs", and $K_I \sqsubseteq C I$ stands for "K_I conceptualizes I", arrows denote subsumption and solid lines denote complementarity (top) and logical negation with respect to the modal formula (bottom)

$$I \text{ viprs:n\_loc\_cor\_rnoc } J \;\equiv\; \exists K_I, K_J \in \mathbf{I} : (K_I \mathbf{C} I) \wedge (K_J \mathbf{C} J)$$
$$\wedge \;\square\big(\exists I', J' : (K_I \mathbf{C} I') \wedge (K_J \mathbf{C} J') \wedge (\text{occ}[J'] \;\rightarrow\; \text{occ}[I'])\big). \quad (5.6)$$

Relations from the EMMO are mereological (or, more properly, mereotopological [10, 18, 19]), represented here at the highest level by proper parthood

$$\mathbf{P} \;\equiv\; \text{viprs:is\_proper\_part\_of} \;\equiv\; \text{emmo-mereotopology:hasProperPart}^-, \quad (5.7)$$

and semiotic, represented at the highest level by the sign-to-object reference relation[6]

$$\mathbf{S} \;\equiv\; \text{viprs:is\_sign\_for} \;\equiv\; \text{emmo-semiotics:hasSign}^-, \quad (5.8)$$

cf. Expressions (3.6) and (3.7). To facilitate ontology alignment, which is discussed in
Sects. 5.3 and 5.4, VIPRS also contains *mereosemiotic* chain products of these funda-
mental relations, i.e. elements of the free semigroup $\mathbf{R}_{ms}^+$ over $\mathbf{R}_{ms} = \{\mathbf{P}, \mathbf{S}, \mathbf{P}^-, \mathbf{S}^-\}$,
with the product defined by concatenation. The mereosemiotic relations for which
there is an explicit definition in VIPRS are limited to $\mathbf{R}_{ms} \cup \mathbf{R}_{ms}^2 \cup \mathbf{R}_{ms}^3$, i.e. relations
generated by a sequence of up to three fundamental relations which are not redundant
($\mathbf{P} \circ \mathbf{P}$ and its inverse),[7] complete (or almost complete), i.e. relating everything to
everything, except possibly for a single "universe" entity,[8] as it is the case for $\mathbf{P} \circ \mathbf{P}^-$,
or consist of three elements from the same category, e.g. $\mathbf{S}^- \circ \mathbf{S} \circ \mathbf{S}^-$ is excluded,
because all three constituent elements are semiotic. In the nomenclature employed

---

[6]It will require an explanation why the shorthand symbols $\mathbf{P}$ ("is proper part of") and $\mathbf{S}$ ("is sign
for") are here assigned a meaning corresponding to the *inverse* of relations defined by the EMMO,
which only include "has proper part" and "has sign," respectively. While making the correspondence
with the EMMO slightly more indirect, we find this to be more in line with common conventions.

First, concerning mereology, parthood is a *partial ordering* relation; such relations are conven-
tionally defined in terms of the operator meaning "is smaller than," not "is greater than," e.g. in
description logic, where subsumption ($\sqsubseteq$) rather than inclusion ($\sqsupseteq$) is employed as a primitive. The
seminal papers on mereotopology by Smith [18], Varzi [10], as well as Smith and Varzi [19] unsur-
prisingly all define "is part of" as fundamental, which they denote by $\mathbf{P}$. We rely on *proper* rather
than improper parthood here based on our empirical assessment that proper parthood is more fre-
quently the most useful choice for ontology alignment, cf. Sect. 5.3, when the EMMO is considered
as a target ontology, cf. Sect. 5.4.

Second, concerning semiotics, when drawing a knowledge graph using the relation "is sign for",
the arrow points from the sign to the object, which is intuitive; with "has sign", the object would
have to point to the sign. In view of this, to avoid a counterintuitive notation that would encourage
the misinterpretation of diagrams, the symbol $\mathbf{S}$ is here employed for "is sign for".

[7]In terms of the 4D spatiotemporal entities considered within the EMMO, $\mathbf{P}$ and $\mathbf{P}^-$ are idempotent,
since for any $I \mathbf{P} J$ there is an I' such that $(I \mathbf{P} I') \wedge (I' \mathbf{P} J)$ due to the continuum nature of
spacetime. The EMMO explicitly permits items to be "void," i.e. not to contain any physical matter,
so that continuum nature can be assumed for EMMO spacetime even concerning properties that are
subject to quantization. Hence, chains that contain $\mathbf{P} \circ \mathbf{P}$ or $\mathbf{P}^- \circ \mathbf{P}^-$ can be excluded.

[8]There is a 4D spatiotemporal entity Ω ("trajectory of the universe") that encloses everything that
exists within any given knowledge base; therefore, "I is a proper part of something (namely, Ω)
that has J as a proper part" holds for all EMMO individuals $I, J \neq \Omega$ from the knowledge base.
Hence, $\mathbf{P} \circ \mathbf{P}^-$ is (almost) complete, and any chains that contain it can be excluded.

by VIPRS, the IRI elements ip, hp, is and hs stand for "is proper part," "has proper part," "is sign" and "has sign," respectively. Accordingly, the binary chain relations include

$$viprs:mereosemiotics\_hp\_ip \equiv P^- \circ P,$$
$$viprs:mereosemiotics\_ip\_is \equiv P \circ S,$$
$$viprs:mereosemiotics\_ip\_hs \equiv P \circ S^-, \text{ etc.} \tag{5.9}$$

while the ternary chain relations include

$$viprs:mereosemiotics\_hp\_ip\_is \equiv P^- \circ P \circ S,$$
$$viprs:mereosemiotics\_hp\_ip\_hs \equiv P^- \circ P \circ S^-,$$
$$viprs:mereosemiotics\_ip\_is\_ip \equiv P \circ S \circ P, \text{ etc.} \tag{5.10}$$

With minor exceptions, datatype properties (owl:DatatypeProperty) are absent from the EMMO [9]; by the domain ontologies, however, datatype properties are amply employed to associate objects with textual (xs:string), numerical (xs:decimal) attributes and xs:boolean flags. Figure 5.4 visualizes the hierarchy of *top-level datatype properties* introduced in VIPRS. At the highest level, VIPRS categorizes datatype properties according to their role:

- Identification of an object is positioned below viprs:has_identifier; examples include otras:has_topic_code, which maps a materials modelling topic (otras:mm_topic) from OTRAS to a four-digit code. Each topic code uniquely corresponds to one topic, and its purpose is identification.
- Where an elementary-datatype entry is the content (or part of the content) of an object, datatype properties below viprs:has_content are used, e.g. this applies to textual or numerical content of MODA from entries (in OSMO, aspects), corresponding to osmo:has_aspect_text_content and osmo:has_aspect_text_content [20, 21], cf. Section 3.3.
- Elementary descriptors, specifiers and similar metadata that provide additional, contingent information on objects, viprs:has_specifier is used, e.g. otras:has_cited_video_duration_seconds points to a metadata item on the length of a video. This contributes to our knowledge about the video by specification, while it does not permit its identification; moreover, the video duration is information about the video content, but it is not itself the content. Therefore, otras:has_cited_video_duration_seconds ⊑ viprs:has_specifier.

At the second level, the datatypes are distinguished (string, decimal or Boolean). Further below, at the third level, the textual datatype properties are further split into subproperties according to their function (cf. Fig. 5.4).

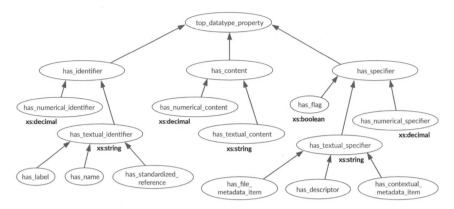

Fig. 5.4 Hierarchy of datatype properties from VIPRS, version 1.0.1; arrows denote subsumption (⊑)

5.3 Ontology Matching

A major design goal for a top-level ontology consists in achieving the desired level of expressivity with a minimal repertoire of basic terms and relations. Obversely, to ensure interoperability for services and tools interoperating at the level of a specific digital platform, the employed ontologies need to capture detailed characteristics of data pertaining to a particular domain of knowledge. Accordingly, the structure of the corresponding semantic space at the lower level is comparably complex, e.g. the ontologies from VIMMP contain about 1000 concepts, 550 relations (object properties) and 180 elementary datatype properties. Therefore, by design, the EMMO needs to have a structure that is substantially different from that of the marketplace-level ontologies [7]. To ensure that the EMMO is consistently employed at all levels, so that it can contribute to platform and service interoperability as far as possible, the marketplace-level ontologies need to be aligned with the EMMO. Before returning to this specific problem, the present section summarizes some of the related theoretical concepts.

In principle, semantic assets are designed to allow data integration and overcome the data heterogeneity problem; in reality, semantic heterogeneity does arise, and it grows over time as resources are added to the semantic web. This is known as the Tower of Babel problem [22, 23]. While some authors regard any presence of semantic heterogeneity as a failure of semantic interoperability and hope for universal agreements, others think that it is unavoidable and look for strategies to deal with it. This may involve a standardized way of documenting semantic assets; basic agreements on the approach to ontology design; and the formalizations of roles, procedures and good practices (or best practices), aiming at pragmatic interoperability [24–27]. For this approach, the challenge consists in agreeing and specifying how the semantic space is structured, documented and employed in practice; by raising the domain for which universal agreements are pursued from the ontological level to the metaontological level, "the Tower of Babel becomes a Meta-Tower of Babel" [28].

As a consequence, semantic heterogeneity is seen as a necessary property of the semantic web, and ontology matching and integration become basic features of its successful mode of operation, rather than an expression of incompleteness. Options for implementing such a mode of operation have been extensively discussed in the literature, first for schemas and then for ontologies, cf. Noy [29] as well as Euzenat and Shvaiko [30]. The common challenge is how to make use of the knowledge represented in two ontologies, which can differ at various levels (language used, expressivity, modelling paradigm, etc.). Typically, such challenges arise if there is an overlap in the domains of knowledge addressed by multiple ontologies, such that data annotated in diverse ways need to be combined and processed together, or if a platform employs multiple domain ontologies that are based on different top-level ontologies. Typical applications include, e.g. simultaneous querying of multiple knowledge bases [31–34] or, as addressed here, the mapping of semantic content from a source ontology S to a target ontology T.

Such a mapping α, by which a scenario A_S expressed in the source ontology is mapped to a A_T expressed in the target ontology, is an *ontology alignment*. Equivalently, this can be applied to the corresponding knowledge graphs, $\alpha : G_S \mapsto G_T$. The process by which an alignment is constructed is known as ontology matching [35]. Alignments can be probabilistic or deterministic, e.g. in a probabilistic formalism, it might be stated that "an osmo:condition that osmo:contains_variable an evmpo:material_property has a 40% probability of being an emmo-models:Physics BasedModel", cf. Suchanek et al. [36]. For the present purpose, we restrict ourselves to deterministic alignments, based on rules that are asserted to be valid in general. If such an alignment is formulated coherently and correctly, the source and target scenarios need to be semantically consistent, i.e. the assertions from the target scenario may not contradict the assertions from the source scenario, which can be checked in multiple ways:

1. Immanently (ontologically), on the basis of a series of alignments $\alpha \circ \alpha' \circ \ldots$, at the end of which another version of the scenario expressed in the source ontology is obtained. Then the consistency of the original and final scenarios can be determined on the basis of the rules from the source ontology S.
2. Transcendentally (metaontologically), either by creating a new ontology that encompasses both S and T, containing rules in which concepts or relations from both ontologies occur jointly, or alternatively by a different system of—possibly human—arbitration that can detect contradictions between A_S and A_T.

Under the constraint of consistency, it is the main challenge to preserve as much of the originally given information as possible. Test scenarios, for which the desired target representation is known, can be used to validate the alignment [34]. Moreover, alignment rules, whether probabilistic or deterministic, can be obtained by evaluating corpora of data that are annotated in both the source and target ontologies [35, 37]; in the probabilistic case, however, the outcome can be assumed to apply only as long as the population or corpus underlying the statistical analysis from which the probabilities were determined is representative of a class of scenarios to which $\mathcal{A}_S$ belongs. Simple alignment correspondences [38] can be specified by categorically subsuming concepts and relations from S under those from T, yielding relabelling rules [39] that do not affect the graph structure (only the labels) and that are context free, i.e. independent of adjacent vertices and edges, such as

$$\text{vivo:evaluates} \sqsubseteq \text{S}, \tag{5.11}$$

stating that whatever evaluates an object, by implication, always also is a sign for that object. Besides, qualified subsumptions can be formulated, such as

$$\exists(\text{vivo:evaluates}^-).\text{evmpo:assertion} \sqsubseteq \text{emmo-semiotics:Object}, \tag{5.12}$$

i.e. that which is evaluated by an assertion is an "object" in the sense of Peircean semiotics; this is a context-sensitive rule, since the relabelling of the vertex (individual) is contingent on one of the edges, namely, an incoming edge with the label vivo:evaluates. Beyond this, more complex graph transformation rules [40] can be applied in the case that the transformation goes beyond relabelling, i.e. if vertices or edges in the knowledge graph need to be eliminated or created by applying $m : n$ property chain correspondences [38].

5.4   VIMMP-EMMO Alignment

To permit the transformation of a knowledge graph from the way in which it appears to the VIMMP marketplace platform to the more abstract representation required for interoperability within a heterogeneous ecosystem of platforms mediated through the EMMO, both concepts and relations need to be aligned between the (VIMMP marketplace) domain level and the (EMMO) top level. This is realized by an ontology module for EMMO-VIMMP Integration (EVI). For the present purpose, accordingly, S is the VIMMP system of ontologies, including the EVMPO (but excluding VIPRS), and T is the EMMO, in the case of concepts, and the EMMO in combination with VIPRS, in the case of relations. In the absence of co-annotated corpora that can be analysed automatically, the correspondences were all specified explicitly, by evaluating the concept and relation definitions from the EMMO alpha version in comparison with the respective definitions from the VIMMP ontologies.

Concerning the conceptual alignment, Fig. 5.5 shows how the categories from the EVMPO, cf. Sect. 3.2, are mapped to EMMO concepts. The red arrows and double lines in Fig. 5.5 represent this alignment, which is itself expressed as an ontology and implemented in the EVI module. This part of the alignment guarantees that all VIMMP domain-ontology concepts are subsumed under EMMO concepts (where they are all situated below emmo-physical:Physical taxonomically), since all of these concepts are either subclasses of one of the fundamental paradigmatic categories from the EVMPO or of the fundamental non-paradigmatic category evmpo:annotation.

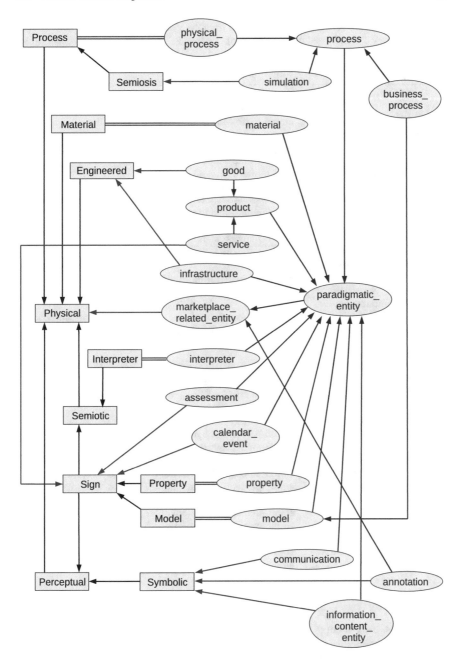

Fig. 5.5 Fundamental categories, superclasses and selected subclasses from the EVMPO (ellipses), version 1.3.1, together with related concepts from EMMO version 1.0.0 alpha 2 (rectangles); arrows between concepts denote subsumption, and double lines between concepts denote equivalence

Table 5.1 Alignment between selected concepts from the VIMMP marketplace-level ontologies (source ontology $\mathcal{S}$) introduced in Sects. 3.3 (top), 3.4 (middle) and 3.5 (bottom) and the EMMO top-level ontology (target ontology $\mathcal{T}$)

source-ontology concept	target-ontology concept
macro:channel	$\sqsubseteq$ emmo-holistic:Process $\sqcap \exists(P^{-})$.emmo-semiotics:Semiosis $\sqcap$ emmo-manufacturing:Engineered
osmo:condition	$\sqsubseteq$ emmo-math:Mathematical
osmo:einecs_listed_material	$\sqsubseteq$ emmo-physicalistic:Material
macro:io_format	$\sqsubseteq$ emmo-perceptual:Symbolic $\sqcap$ emmo-perceptual:Language $\sqcap$ emmo-semiotics:Conventional
osmo:logical_variable	$\equiv$ emmo-math:Variable
osmo:materials_relation	$\equiv$ emmo-models:MaterialRelation
macro:model_database	$\sqsubseteq$ emmo-semiotics:Interpreter $\sqcap$ emmo-manufacturing:Engineered
osmo:section_aspect	$\sqsubseteq$ emmo-physical:Physical
osmo:workflow_graph	$\sqsubseteq$ emmo-perceptual:WellFormedSymbolic
vivo:assertion	$\sqsubseteq$ emmo-perceptual:Symbolic
mmto:business_decision_-_support_system	$\sqsubseteq$ emmo-manufacturing:Engineered
vivo:certificate	$\sqsubseteq$ emmo-perceptual:Symbolic
vivo:computational_-_time_requirement	$\sqsubseteq \exists S.$(emmo-semiotics:Sign $\sqcap$ emmo-perceptual:Symbolic) $\sqcap \exists(S \circ S)$.emmo-semiotics:Semiosis $\sqcap$ emmo-semiotics:Sign
vivo:material_property_-_information	$\sqsubseteq$ emmo-semiotics:Sign $\sqcap \exists S$.emmo-properties:Property $\sqcap \exists S$.emmo-physicalistic:Material $\sqcap \exists(S \circ S)$.emmo-physicalistic:Material
mmto:objective_kpi	$\sqsubseteq$ emmo-properties:ObjectiveProperty
vivo:relative_deviation	$\sqsubseteq$ emmo-math:Mathematical
mmto:translation_case	$\sqsubseteq$ emmo-perceptual:Symbolic $\sqcap$ emmo-semiotics:Sign
vivo:unit	$\equiv$ emmo-metrology:ReferenceUnit
vico:academic_title	$\sqsubseteq$ emmo-perceptual:Symbolic $\sqcap \exists P$.emmo-semiotics:Sign
otras:focus	$\sqsubseteq$ emmo-perceptual:Symbolic $\sqcap \exists P$.emmo-semiotics:Sign
vico:interlocutor_tag	$\sqsubseteq$ emmo-perceptual:Symbolic $\sqcap \exists P$.emmo-semiotics:Sign
vico:message	$\sqsubseteq$ emmo-perceptual:Symbolic
otras:mm_topic	$\sqsubseteq$ emmo-perceptual:Symbolic $\sqcap$ emmo-semiotics:Sign
vico:person	$\sqsubseteq$ emmo-semiotics:Interpreter
otras:specifier	$\sqsubseteq$ emmo-perceptual:Symbolic $\sqcap \exists P$.emmo-semiotics:Sign
otras:training_service	$\sqsubseteq$ emmo-semiotics:Sign $\sqcap \exists S$.emmo-holistic:Process
otras:training_unit	$\sqsubseteq$ emmo-holistic:Process

Beyond this, the concepts from the domain ontologies are aligned with the EMMO down to a comparably fine-grained level; this is also implemented in EVI.[9] Table 5.1 contains the EVI statements corresponding to the concepts that were listed as examples in Chap. 3.

---

[9]EVI: `https://purl.vimmp.eu/semantics/alignment/evi.ttl` (non-resolvable IRI), mirrored at `http://www.molmod.info/semantics/evi.ttl` (resolvable URL).

Table 5.2 Alignment between selected relations from the VIMMP marketplace-level ontologies (source ontology S) introduced in Sects. 3.3 (top), 3.4 (middle) and 3.5 (bottom) and VIPRS in combination with the EMMO (target ontology T)

source-ontology relation	target-ontology relation[a,b,c,d]
osmo:has_aspect $\sqsubseteq$	$S^- \circ (P^- \sqcap {}^\bullet$emmo-perceptual:Symbolic$)$
osmo:has_aspect_object_content $\sqsubseteq$	$P^- \circ S$
macro:has_channel_member $\sqsubseteq$	emmo-holistic:hasParticipant
macro:has_granularity $\sqsubseteq$	$(S^- \circ P^-) \sqcap (S \circ P^- \circ S^-)$
osmo:has_value $\sqsubseteq$	$S^- \circ (P^- \sqcap {}^\bullet$emmo-perceptual:Symbolic$)$
osmo:has_variable_unit $\sqsubseteq$	emmo-metrology:hasReferenceUnit
macro:is_io_format_of $\sqsubseteq$	$S \circ$ viprs:p_loc_and_roc, *cf.* note a
osmo:is_linked_to $\sqsubseteq$	viprs:mutual_requirement, *cf.* notes b and c
macro:provides_access_to $\sqsubseteq$	$(S \circ P^-) \sqcap$ viprs:satisfies_requirement_of, *cf.* note c
mmto:considers_business_case $\sqsubseteq$	$P^- \circ P \circ S^-$
mmto:describes_product $\sqsubseteq$	S
vivo:evaluates $\sqsubseteq$	S
vivo:has_assertion $\sqsubseteq$	P^-
vivo:has_error_statement $\sqsubseteq$	P^-
mmto:has_tca_pe_type $\sqsubseteq$	$S^- \circ (P^- \sqcap {}^\bullet$emmo-perceptual:Symbolic$)$
vivo:has_unit $\sqsubseteq$	$S^- \circ (P^- \sqcap {}^\bullet$emmo-perceptual:Symbolic$)$
vivo:is_quantity_kind $\sqsubseteq$	S^-
vivo:states_assessment $\sqsubseteq$	$P^- \sqcap$ viprs:enables, *cf.* note d
vico:contains $\sqsubseteq$	P^-
vico:follows $\sqsubseteq$	$S \sqcap$ viprs:is_enabled_by, *cf.* note d
vico:has_affiliation $\sqsubseteq$	P
vico:has_author $\sqsubseteq$	viprs:is_enabled_by, *cf.* note d
otras:has_offered_course $\sqsubseteq$	$S \circ P^-$
otras:has_specifier $\sqsubseteq$	$S^- \circ P^-$
vico:is_certifier_of $\sqsubseteq$	viprs:satisfies_requirement_of, *cf.* note c
otras:is_narrower_than $\sqsubseteq$	$S \circ S^-$
otras:is_part_of_course $\sqsubseteq$	P

[a]Modal statement I viprs:p_loc_and_roc J (read p_loc_and_roc as "possibly, the left occurs and the right occurs"): It is possible for the left-hand argument I and the right-hand argument J to occur jointly, i.e. $\Diamond(\text{occ}[I] \wedge \text{occ}[J])$.

[b]The modal relation viprs:mutual_requirement is defined by equivalence with the relational intersection
viprs:satisfies_ requirement_of $\sqcap$ viprs:satisfies_requirement_of$^-$.

[c]I viprs:satisfies_requirement_of J implies that for some conceptualization K_I of I, it is necessary that an instantiation I' of K_I occurs or that J does not occur

$$I \text{ viprs:satisfies\_requirement\_of } J \implies \tag{5.13}$$
$$\exists K_I : K_I \mathbf{C} I \wedge \Box(\exists I' : (K_I \mathbf{C} I') \wedge (\text{occ}[J] \to \text{occ}[I']));$$

e.g. if J is a simulation with a code that requires XML input, I might be a particular XML file, satisfying a requirement of J, while K_I might be the XML file format. Then it is not the occurrence of I itself that is strictly required for J to occur, but the occurrence of the more generic object K_I. However, $\text{occ}[I]$ necessarily implies $\text{occ}[K_I]$. By comparing Eqs. (5.6) and (5.13), with $K_J = J$, we observe

$$\text{viprs:satisfies\_requirement\_of} \sqsubseteq \text{viprs:n\_loc\_cor\_rnoc}. \tag{5.14}$$

[d]The modal relation viprs:is_enabled_by is defined by equivalence with the relational intersection viprs:p_loc_and_roc $\sqcap$ viprs:satisfies_requirement_of$^-$, the inverse relation of which is given by viprs:enables $\equiv$ viprs:p_loc_and_roc $\sqcap$ viprs:satisfies_requirement_of

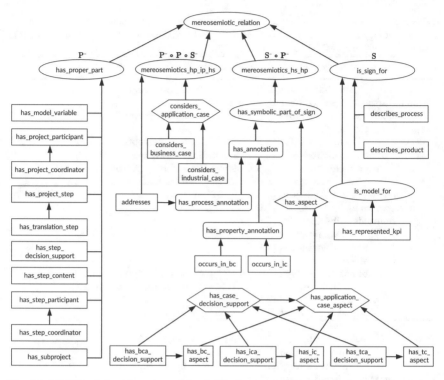

Fig. 5.6 MMTO relation hierarchy, version 1.3.1, showing the subsumption (arrows) of relations from the MMTO (rectangles) under relations from OSMO (hexagons), the EVMPO (rounded boxes) and VIPRS (ellipses)

The relational alignment, which is shown for the MMTO in Fig. 5.6 and for the examples from Chap. 3 in Table 5.2, is implemented directly in the domain ontology TTL files, which contain statements by which the domain ontology relations are subsumed under VIPRS relations. Property chain correspondences are applied when the mereosemiotic chain relations from VIPRS are unfolded, cf. Fig. 5.7, yielding series of elementary parthood and reference relations from the EMMO, so that the graph grows both in terms of vertices and edges; in TTL notation, this corresponds to the introduction of blank nodes (individuals without an IRI [41]) by which, e.g.

```
:I viprs:mereosemiotics_hp_ip_is :J.
```

which encodes (:l, :J) : $\left(\mathsf{P}^- \circ \mathsf{P} \circ \mathsf{S} \right)$ becomes

```
:J emmo-semiotics:has_sign [
    emmo-mereotopology:has_proper_part [
        [owl:inverseOf emmo-mereotopology:has_proper_part] :I
    ]
].
```

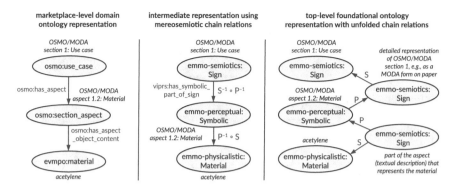

Fig. 5.7 Correspondences between the domain and top levels; example: Description of a materials modelling use case following MODA [20] and OSMO [21]. Ellipses denote individuals, labelled by the concept names from the respective ontologies (EVMPO, OSMO and multiple EMMO modules), and arrows denote relations. At the intermediate stage, mereosemiotic chain relations from VIPRS are used to support the alignment [42]

5.5 Documentation of Molecular Models

For documenting molecular models and exchanging them between platforms, a semantic interoperability standard on the basis of the VIMMP system of ontologies as well as MODA [20] was agreed between VIMMP and the Molecular Model Database (MolMod DB) of the Boltzmann-Zuse Society [43]; the associated environment of interoperable platforms will prospectively also include Bottled SAFT [44].

The structure of the knowledge graph representing a molecular model is illustrated by Fig. 5.8, which corresponds to a two-centre Lennard-Jones plus point-quadrupole model (2CLJQ) where a molecule (in this case, acetylene) is represented by a rigid unit (viso-am:rigid_object), consisting of two Lennard-Jones interaction sites (viso-am:lj_site), a point-quadrupole site (viso-am:charge_quadrupole_site) as well as a viso-am:structureless_object, representing the molecular centre of mass, which is used the initial point of vov:relative_position vectors that indicate the coordinates of the interaction sites. The relation vov:has_attached_variable and subproperties of it are used to connect the interaction sites with the non-geometrical model parameters, i.e. the mass associated with each of the two LJ sites (half the molecular mass), the σ and ϵ site and energy parameters of the LJ potential, and a second-order tensor characterizing the quadrupole moment. Other rigid molecular models are described analogously.

The platform interoperability implementation developed on this basis employs JSON-LD to exchange information on molecular models. Therefore, the knowledge graph needs to be connected (i.e. it may not consist of multiple connected components), and its topology needs to be simplified to a tree structure such that each object is subordinate to exactly one object, except for a single root node at the top. For the present example, an osmo:workflow_graph with two sections, a use case (MODA

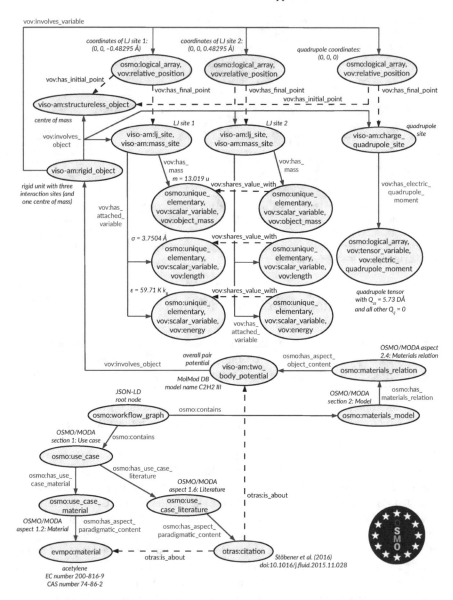

Fig. 5.8 Knowledge graph representing a 2CLJQ model for acetylene by Stöbener et al. [45], i.e. model ID 97 (C$_2$H$_2$ III) from the MolMod DB [43]. Ellipses denote individuals, labelled by the concept names from the respective ontologies (EVMPO, OSMO, OTRAS, VISO and VOV) and arrows denote relations. The graph from Fig. 5.7 is included in the bottom left corner [42]

Sect. 5.1) and a model (MODA Sect. 5.2), is selected as the root of the tree. In this way, e.g. one of the site coordinates is included in the rigid unit description as follows:

```
{
    "@id": "acetylene:RIGID_UNIT",
    "@type": "viso-am:rigid_object",
    "vov:involves_object": [
        {
            "@id": "acetylene:COM",
            "@type": "viso-am:structureless_object"
        }, {
            "@id": "acetylene:SITE_LJ_A",
            "@type": ["viso-am:lj_site", "viso-am:mass_site"], ...
        }, ...
    ],
    "vov:involves_variable": [
        {
            "@id": "acetylene:LJ_A_POS",
            "@type": ["osmo:logical_array", "vov:relative_position"],
            "vov:has_initial_point": {"@id": "acetylene:COM"},
            "vov:has_final_point": {"@id": "acetylene:SITE_LJ_A"}, ...
        }, ...
    ]
}
```

The hierarchy by which objects are embedded in other objects in JSON is obtained from a subset of the relations from the knowledge graph, shown in Fig. 5.8 as solid arrows in blue colour, while references to IRIs are used to represent the other relations (dashed black arrows) in JSON-LD. The relations vov:involves_object and vov:involves_variable are part of the JSON-LD tree structure (solid blue arrows), so that COM, SITE_LJ_A, and LJ_A_POS are all hierarchically subordinate to RIGID_UNIT. The relations vov:has_initial_point and vov:has_final_point, however, are sideways connections between nodes from multiple branches of the tree (dashed black arrows); therefore, their JSON-LD representation only points to the IRI of the referenced object, using the `"@id"` keyword.

References

1. F. Abad Navarro, J.A. Bernabé Diaz, A. García Castro, J.T. Fernández Breis, Semantic publication of agricultural scientific literature using property graphs. Appl. Sci. **10**(3), 861 (2020)
2. E. Katis, H. Kondylakis, G. Agathangelos, V. Kostas, Developing an ontology for curriculum & syllabus, in *Prof. ESWC, Satellite Events*, ed. by A. Gangemi, A.L. Gentile, A.G. Nuzzolese, M. Rudolph, S. Maleshkova, H. Paulheim, J.Z. Pan, M. Alam (Springer, Cham, Switzerland, 2018), pp. 55–59
3. G.J. Schmitz, Microstructure modeling in integrated computational materials engineering (ICME) settings: can HDF5 provide the basis for an emerging standard for describing microstructures? JOM **68**(1), 77–83 (2016)

4. P. de Buyl, P.H. Colberg, F. Höfling, H5MD: a structured, efficient, and portable file format for molecular data. Comput. Phys. Commun. **185**, 1546–1553 (2014)
5. W. Colsman, R. Uphill, *A portable data format for laboratory data* (Sci. Comput, World Feature, 2015)
6. H. Oberkampf, H. Krieg, C. Senger, T. Weber, W. Colsman, Allotrope data format: semantic data management in life sciences, in *Proceedings of SWAT4HCLS 2018*, ed. by A. Splendani (2018)
7. M.T. Horsch, S. Chiacchiera, M.A. Seaton, I.T. Todorov, K. Šindelka, M. Lísal, B. Andreon, E.B. Kaiser, G. Mogni, G. Goldbeck, R. Kunze, G. Summer, A. Fiseni, H. Brüning, P. Schiffels, W.L. Cavalcanti, Ontologies for the Virtual Materials Marketplace. Künstl. Intell. **34**(3), 423–428 (2020). https://doi.org/10.1007/s13218-020-00648-9
8. G. Goldbeck, E. Ghedini, A. Hashibon, G.J. Schmitz, J. Friis, A reference language and ontology for materials modelling and interoperability, in *Proceedings of NWC 2019*, NAFEMS, (Knutsford, UK, 2019), p NWC_19_86
9. EMMC Coordination and Support Action, European Materials and Modelling Ontology (2020), https://github.com/emmo-repo/, https://emmc.info/emmo-info/. Accessed 8 Apr 2020
10. A.C. Varzi, Parts, wholes, and part-whole relations: the prospects of mereotopology. Data Knowl. Eng. **20**, 259–286 (1996)
11. C.S. Peirce, *Peirce on Signs: Writings on Semiotic* (University of North Carolina Press, Chapel Hill, North Carolina, USA, 1991)
12. D. Lewis, New work for a theory of universals. Aust. J. Philos. **61**(4), 343–377 (1983)
13. M.T. Horsch, S. Chiacchiera, B. Schembera, M.A. Seaton, I.T. Todorov, Semantic interoperability based on the European materials and modelling ontology and its ontological paradigm: mereosemiotics, in *Proceedings of WCCM-ECCOMAS 2020*, to appear (2021). https://doi.org/10.5281/zenodo.3902900
14. D. Westerståhl, The traditional square of opposition and generalized quantifiers. Stud. Logic (Beijing) **2**, 1–18 (2008)
15. D. Westerståhl, Classical vs. modern squares of opposition, and beyond, in *The Square of Opposition: A General Framework for Cognition*, ed. by J.Y. Béziau, G. Payette (Switzerland, Peter Lang, Bern, 2012), pp. 195–229
16. F. Berto, M. Plebani, *Ontology and Metaontology* (Bloomsbury, London, UK, 2015)
17. M. Huth, M. Ryan, *Logic in Computer Science: Modelling and Reasoning about Systems*, 2nd edn. (Cambridge University Press, Cambridge, 2004)
18. B. Smith, Mereotopology: a theory of parts and boundaries. Data Knowl. Eng. **20**(3), 287–303 (1996)
19. B. Smith, A.C. Varzi, Fiat and bona fide boundaries. Philos. Phenomenol. Res. **60**(2), 103–119 (2000)
20. CEN-CENELEC Management Centre, *Materials modelling: terminology, classification and metadata, in CEN Workshop Agreement 17284* (Belgium, Brussels, 2018)
21. M.T. Horsch, C. Niethammer, G. Boccardo, P. Carbone, S. Chiacchiera, M. Chiricotto, J.D. Elliott, V. Lobaskin, P. Neumann, P. Schiffels, M.A. Seaton, I.T. Todorov, J. Vrabec, W.L. Cavalcanti, Semantic interoperability and characterization of data provenance in computational molecular engineering. J. Chem. Eng. Data **65**(3), 1313–1329 (2020)
22. B. Hu, B. Hu, Tower of Babel: Interoperability of ontologies for pervasive computing, in *First International Symposium on Pervasive Computing and Applications*, ed. by V. Callaghan, B. Hu, Z. Lin, H. Zhang (IEEE, Piscataway, New Jersey, USA, 2006), pp. 690–695
23. A. Iliadis, The tower of Babel problem: making data make sense with basic formal ontology. Online Inf. Rev. **43**(6), 1021–1045 (2019)
24. C.H. Asuncion, M.J. van Sunderen, Pragmatic interoperability: a systematic review of published definitions, in *Proceedings of EAI2N, WCC 2010*, ed. by P. Bernus, G. Doumeingts, M. Fox (Springer, Heidelberg, Germany, 2010), pp. 164–175
25. M.T. Horsch, S. Chiacchiera, M.A. Seaton, I.T. Todorov, B. Schembera, P. Klein, N.A. Konchakova, Pragmatic interoperability and translation of industrial engineering problems into modelling and simulation solutions, in *Proceedings of DAMDID 2020*, to appear (2021). https://doi.org/10.5281/zenodo.3902873

26. B. Schembera, J.M. Durán, Dark data as the new challenge for big data science and the introduction of the scientific data officer. Philos. Technol. **33**, 93–115 (2020)
27. M. Schoop, A. de Moor, J. Dietz, The pragmatic web: a manifesto. Commun. ACM **49**(5), 75–76 (2006)
28. Varzi A (2019) Ontology: from philosophy to innovation in materials and manufacturing. Keynote address, in *2nd EU Workshop on Materials and Manufacturing Ontology*, Brussels, 6th June 2019
29. N.F. Noy, Semantic integration: a survey of ontology-based approaches. SIGMOD Rec. **33**(4), 65–70 (2004)
30. J. Euzenat, P. Shvaiko, *Ontology Matching*, 2nd edn. (Springer, Heidelberg, 2013)
31. B. Bouchou, C. Niang, Semantic mediator querying, in *Proceedings of IDEAS '14*, ed. by A.M. Almeida, J. Bernardino, E. Ferreira Gomes (ACM, New York, USA, 2014), pp 29–38
32. D. Lembo, R. Rosati, V. Santarelli, D.F. Savo, E. Thorstensen, (2017) Mapping repair in ontology-based data access evolving systems, in *Proceedings of IJCAI*, IJCAI, ed. by C. Sierra (San José, California, USA, 2017), pp. 1160–1166
33. G. Xiao, D. Calvanese, R. Kontchakov, D. Lembo, A. Poggi, R. Rosati, M. Zakharyaschev, (2018) Ontology-based data access: a survey, in *Proceedings of IJCAI*, IJCAI, ed. by J. Lang, (San José, California, USA, 2018), pp. 5511–5519
34. G. Fusco, L. Aversano, An approach for semantic integration of heterogeneous data sources. PeerJ Comput. Sci. **6**, e254 (2020). https://doi.org/10.7717/peerj-cs.254
35. P. Ochieng, S. Kyanda, A statistically-based ontology matching tool. Distrib. Parall. Databases **36**, 195–217 (2018)
36. F.M. Suchanek, S. Abiteboul, P. Senellart, PARIS: probabilistic alignment of relations, instances, and schema. Proc. VLDB Endow. **5**(3), 157–168 (2011)
37. M. Koutraki, N. Preda, D. Vodislav, Online relation alignment for linked datasets, in *Proceedings of ESWC 2017*, ed. by E. Blomqvist, D. Maynard, A. Gangemi, R. Hoekstra, P. Hitzler, O. Hartig (Springer, Cham, Switzerland, 2017), pp. 152–168
38. L. Zhou, M. Cheatham, P. Hitzler, Towards association rule-based complex ontology alignment, in *Proceedings of JIST 2019*, ed. by X. Wang, F.A. Lisi, G. Xiao, E. Botoeva, LNCS, vol. 10249 (Springer, Cham, Switzerland, 2020), pp. 287–303
39. Y. Métivier, E. Sopena, Graph relabelling systems: a general overview. Comput. AI **16**(2), 167–185 (1997)
40. B. König, D. Nolte, J. Padberg, A. Rensink, A tutorial on graph transformation, in *Graph Transformation, Specifications, and Nets*, ed. by R. Heckel, G. Taentzer, LNCS, vol. 12032 (Springer, Cham, 2018), pp. 83–104
41. D. Allemang, J. Hendler, *Semantic Web for the Working Ontologist*, 2nd edn. (Morgan Kaufmann, Waltham, Massachusetts, USA, 2011)
42. M.T. Horsch, S. Chiacchiera, M.A. Seaton, I.T. Todorov, R. Kunze, G. Summer, A. Fiseni, B. Andreon, A. Scotto Di Minico, E. Bayro Kaiser, G. Kanagalingam, S. Stephan, K.Šindelka, M. Lísal, J. Díaz Brañas, I. Pagonabarraga, M. Chiricotto, J.D. Elliott, P. Carbone, D. Toti, G. Mogni, G. Goldbeck, H. Brüning, P. Schiffels, W.L. Cavalcanti, Ontology-based semantic interoperability on the virtual materials marketplace, in *Proceedings of the ISWC 2020 Demos and Industry Tracks*, ed. by K. Taylor, R. Gonçalves, F. Lecue, J. Yan (CEUR-WS, Aachen, 2021), pp. 134–137. https://doi.org/10.5281/zenodo.3986825
43. S. Stephan, M.T. Horsch, J. Vrabec, H. Hasse, MolMod - an open access database of force fields for molecular simulations of fluids. Mol. Sim. **45**(10), 806–814 (2019)
44. Å. Ervik, A. Mejía, E.A. Müller, Bottled SAFT: a web app providing SAFT-γ Mie force field parameters for thousands of molecular fluids. J. Chem. Inf. Model. **56**(9), 1609–1614 (2016)
45. K. Stöbener, P. Klein, M. Horsch, K. Küfer, H. Hasse, Parametrization of two-center Lennard-Jones plus point-quadrupole force field models by multicriteria optimization. Fluid Phase Equilib. **411**, 33–42 (2016)

Printed in the United States
by Baker & Taylor Publisher Services